The NASA STI Program Office . . . in Profile

Since its founding, NASA has been dedicated to the advancement of aeronautics and space science. The NASA Scientific and Technical Information (STI) Program Office plays a key part in helping NASA maintain this important role.

The NASA STI Program Office is operated by Langley Research Center, the Lead Center for NASA's scientific and technical information. The NASA STI Program Office provides access to the NASA STI Database, the largest collection of aeronautical and space science STI in the world. The Program Office is also NASA's institutional mechanism for disseminating the results of its research and development activities. These results are published by NASA in the NASA STI Report Series, which includes the following report types:

- TECHNICAL PUBLICATION. Reports of completed research or a major significant phase of research that present the results of NASA programs and include extensive data or theoretical analysis. Includes compilations of significant scientific and technical data and information deemed to be of continuing reference value. NASA's counterpart of peer-reviewed formal professional papers but has less stringent limitations on manuscript length and extent of graphic presentations.

- TECHNICAL MEMORANDUM. Scientific and technical findings that are preliminary or of specialized interest, e.g., quick release reports, working papers, and bibliographies that contain minimal annotation. Does not contain extensive analysis.

- CONTRACTOR REPORT. Scientific and technical findings by NASA-sponsored contractors and grantees.

- CONFERENCE PUBLICATION. Collected papers from scientific and technical conferences, symposia, seminars, or other meetings sponsored or cosponsored by NASA.

- SPECIAL PUBLICATION. Scientific, technical, or historical information from NASA programs, projects, and missions, often concerned with subjects having substantial public interest.

- TECHNICAL TRANSLATION. English-language translations of foreign scientific and technical material pertinent to NASA's mission.

Specialized services that complement the STI Program Office's diverse offerings include creating custom thesauri, building customized databases, organizing and publishing research results . . . even providing videos.

For more information about the NASA STI Program Office, see the following:

- Access the NASA STI Program Home Page at *http://www.sti.nasa.gov*

- E-mail your question via the Internet to help@sti.nasa.gov

- Fax your question to the NASA Access Help Desk at (301) 621-0134

- Telephone the NASA Access Help Desk at (301) 621-0390

- Write to:
  NASA Access Help Desk
  NASA Center for AeroSpace Information
  7121 Standard Drive
  Hanover, MD 21076-1320

NASA/TP–2006-213486

Notes on Earth Atmospheric Entry for Mars Sample Return Missions

Thomas Rivell
Ames Research Center, Moffett Field, California

National Aeronautics and
Space Administration

Ames Research Center
Moffett Field, California 94035-1000

September 2006

ACKNOWLEDGMENTS

The author would like to thank the following people for their advice, comments, suggestions and encouragement:

Michael L. Hines envisioned the need to introduce newcomers to the subject of atmospheric entry for sample return missions. He proposed the project and provided time to complete it.

Timothy P. Castellano provided encouragement and assistance. He served as my sponsor as an Ames Associate after my retirement from NASA.

Michael E. Tauber, a friend and former college classmate, read the manuscript and provided many valuable suggestions.

Gary A. Allen, Jr. read the manuscript and provided many valuable suggestions. Allen also provided the data for some of the figures.

Roger Arno and Paul Kolodziej read the manuscript and provided many helpful comments and suggestions.

Marcus Murbach contributed some material on flight testing and deceleration and recovery of entry vehicles.

Lawrence E. Lasher and Lisa Chu-Thielbar read the manuscript and provided helpful suggestions.

Jane Parham read an earlier version of the manuscript and tried to make it consistent and readable.

Daniel Pappas, a NASA reference librarian, was extremely helpful in locating books and reports as well as performing literature searches.

Note that any mistakes, errors, inaccuracies or miscalculations are the fault of the author.

Available from:

NASA Center for AeroSpace Information	National Technical Information Service
7121 Standard Drive	5285 Port Royal Road
Hanover, MD 21076-1320	Springfield, VA 22161
(301) 621-0390	(703) 487-4650

TABLE OF CONTENTS

LIST OF FIGURES

LIST OF TABLES

PREFACE

In the second decade of the 21st century, NASA plans additional science orbiters, rovers, and landers. One proposal is for a Mars sample return mission that would use robotic systems and a Mars ascent rocket to collect and send samples of Martian rocks, soils, and atmosphere to Earth for detailed chemical and physical analysis. Researchers on Earth could measure chemical and physical characteristics much more precisely than they could by remote control.

Jet Propulsion Laboratory, Mars Exploration Website, 2006

There has been interest in obtaining gas, dust, rocks, or soil samples from comets or nearby planets since the beginning of the space age in the 1950s. Flying a spacecraft to the destination and returning a sample in an Earth entry vehicle can accomplish this. The return spacecraft is called a sample return vehicle (SRV). Detailed studies of Martian samples can only be conducted on Earth.

In the past decade there have been many studies investigating the feasibility and practicalities of sample return missions to Mars; see Wercinski, 1996; Mitcheltree et al., 1998; and Desai et al., 2000. Samples collected on Mars will be sealed in a special capsule to prevent contamination. An ascent vehicle will launch the capsule from the surface of Mars. There are several ways to return the sample to Earth. One way is to use a Mars orbiter to capture the sample container and then begin an interplanetary trajectory to return the sample to Earth.

This document is concerned with the Earth entry environment and the Earth entry portion of a sample return mission to the planet Mars. The document is intended to introduce the subject of atmospheric entry to engineers and scientists who do not have strong backgrounds in aerodynamics, aerothermodynamics and flight mechanics. Much of this material can be applied to entry into other planetary atmospheres by using the appropriate atmospheric and gravity models. This document discusses topics in atmospheric entry and presents examples in related disciplines. References are cited for further information. The document is not intended to be comprehensive and some important topics are omitted (e.g., chemically reacting gases and nonequilibrium flows). There is an extremely large number of engineering and scientific books, journal articles, conference papers, and government, industry and university reports on the subject of atmospheric entry (see National Aeronautics and Space Administration (NASA), American Institute of Aeronautics and Astronautics (AIAA), Department of Defense, and commercial online databases for several thousand citations on atmospheric entry and/or re-entry vehicles).

The topics considered in this document include basic principles of physics (fluid mechanics, rigid body dynamics, and heat transfer), chemistry, and engineering. The following subjects are discussed:

Fluid mechanics – Aerodynamics, compressible fluids, shock waves, continuum flow (incompressible to hypersonic), hypervelocity flow (high temperature gas dynamics), boundary layers and viscosity, and free molecular flow.

Earth's atmosphere and gravity – Earth model atmosphere and acceleration due to gravity.

Heat and mass transfer – Forced convection, radiation, and ablation.

Dynamics – Rigid body dynamics, and static and dynamic stability.

Flight and ground test simulation – Flight simulation and ground test facility capabilities.

Numerical simulation – A brief description of computational fluid dynamics (CFD) and trajectory simulation tools.

For definitions of unfamiliar terms see Definition of Technical Terms for Aerospace Use (Allen, 1965), or the current online version (Glover, 2006).

A brief historical note may help to introduce the material that follows. In October 1957 the Soviet Union launched the first artificial satellite, Sputnik I, and the space age began. In 1958 the U.S. Congress passed the Space Act creating the National Aeronautics and Space Administration (NASA). In 1961, President John Kennedy announced that the U.S. would land a man on the moon and return him safely to Earth before the end of the decade, a feat accomplished by the Apollo 11 mission in July 1969.

In the 1960s the U.S. and the Soviet Union developed intercontinental ballistic missiles (ICBMs) capable of flying more than 10,000 km in 30 minutes. The U.S. solid propellant Minuteman II was operational in the 1960s and was capable of a range of 11,000 km. The Minuteman II weighed 31,750 kg at liftoff and traveled at 6.7 km/sec at third-stage burnout. The two-stage, liquid propellant Titan II, also operational in the 1960s, was capable of a range of 15,000 km and reached an altitude of 1500 km. The burnout velocity for the Titan II was greater than 6.7 km/sec and liftoff weight was 149,700 kg (Blake, 1988). Atlas, Titan and Minuteman ICBMs have been used as launch vehicles for satellites and spacecraft. By the 1970s Soviet ICBMs were capable of 16,000-km ranges with liquid propellant rockets.

Prior to 1958, the National Advisory Committee for Aeronautics (NACA), the predecessor of NASA, was involved in atmospheric-entry research. NACA engineers and scientists experimentally and analytically studied supersonic and hypersonic flows. The "blunt body concept" to reduce aerodynamic entry heating was formulated by NACA's H. J. Allen in the 1950s (Allen and Eggers, 1958). Allen demonstrated that blunt, high-drag bodies are superior to pointed, slender bodies in atmospheric entry. This is due to the fact that the stagnation-point, convective heat flux decreases as bluntness increases. The higher drag of the blunt body also tends to reduce the convective heat flux by causing the body to decelerate at higher altitudes (see sec. 4.2). The Mercury, Gemini and Apollo entry capsules were applications of the blunt body concept. The early days of the space age and the formation of NASA are described in a recent NASA monograph (Portree, 1998).

Lander spacecraft are designed to reach the surface of a planet and survive long enough to record and transmit scientific data. The first successful planetary landing was by the Soviet Union on Venus in 1966. The Soviet Venera landers survived the harsh conditions on Venus while carrying out chemical composition analyses of the rocks and relaying color images. The Soviets also landed spacecraft on Mars in 1974. The United States' Mars Viking landers in 1976 and the earlier Surveyor landers on the moon carried out similar experiments. NASA participated extensively in the design and development of many successful planetary probes including the Viking Mars lander, Pioneer Venus, Galileo (Jupiter arrival in 1995) and Mars Pathfinder in 1997.

The Space Transportation System (STS), or Space Shuttle Orbiter, was the subject of extensive ground testing and numerical simulation. An extremely important contribution to the Orbiter's success was the NASA Ames-Lockheed development of a revolutionary thermal protection system (TPS). Lockheed furnished lightweight ceramic tiles and Ames provided tile coatings and flexible blankets to shield the vehicle and crew from the intense entry heat loads. Similar TPS technology developments made possible many successful interplanetary missions. There have also been remarkable advances in computational fluid mechanics and in information technology in the past decade. These facts are mentioned because they will have a significant influence on future space exploration.

This document is an attempt to describe some of the basic concepts in aerodynamics and aerothermodynamics that relate to the entry of a Mars sample return vehicle (SRV) into the Earth's atmosphere.

A personal note – When the first Sputnik was launched I was a graduate student in engineering. In the years that followed there were many changes in the engineering curriculum. Engineering colleges began to offer courses in celestial mechanics and "space technology." University departments of aeronautics and mechanical engineering added aerospace engineering and astronautics to the curriculum. More emphasis was placed on hypervelocity flight, rarefied (low density or free molecular) flow and high temperature gas dynamics. Journals, conference papers, courses and textbooks were devoted to these subjects. Universities and industry began to place greater emphasis on collaboration between scientists and engineers.

Note on Internet resources – There is a great deal of useful information on aeronautics and space on the Internet (World Wide Web). Unfortunately, websites change frequently and often disappear. Several websites are mentioned in this document and listed in the References and Bibliography sections. Most of the sites listed are those of professional societies and other well-established organizations.

NASA Photo No. A-22664 (1957)

H. Julian Allen is best known for the "Blunt Body Theory," an aerodynamic design technique for alleviating the severe re-entry heating, which was delaying the development of ballistic missiles. His findings revolutionized the fundamental design of ballistic missile re-entry vehicles. Subsequent research led to the use of blunt shapes for ballistic missiles and spacecraft re-entering the Earth's atmosphere. Blunt, ablative heat shields were used to protect the Mercury, Gemini and Apollo astronauts as their space capsules returned to Earth.

1. INTRODUCTION

1.1 Scope

This document is intended to provide an introduction to Earth atmospheric entry for the nonspecialist. However, a basic knowledge of engineering or physics is required. The purpose of this work is to describe various aerodynamic, aerothermodynamic, and flight-dynamic aspects of the high-speed entry of sample return vehicles (SRVs) into the Earth's atmosphere. SRVs returning from missions to Mars would have entry velocities of 11 to 15 km/s.[1] Flight velocities of this magnitude are often referred to as hypervelocity rather than hypersonic. In hypervelocity flight high-temperature chemical reactions significantly affect the airflow around an entry vehicle. At altitudes above 80 km the density of the atmosphere is so low that the average distance an air molecule travels between successive collisions with other air molecules is large compared to the size of the SRV. This condition is referred to as the free molecular flow regime in contrast to the continuum flow regime at lower altitudes.

Aerodynamic and gravitational forces strongly influence SRV entry dynamics and heat transfer characteristics. Knowledge of atmospheric properties is necessary to determine aerodynamic and heating loads. Aerodynamic forces and heat transfer characteristics are functions of atmospheric density and vehicle speed. A model Earth atmosphere is discussed briefly in section 2.

The design an SRV for a specific mission requires knowledge of the characteristics of the flow around the vehicle. Detailed flow field information may not be necessary if experimental data are available for similar vehicles and flight conditions. Computational fluid dynamics (CFD)[2] solutions that have been validated can also be used in the vehicle design process.

Preliminary trajectory calculations can be used to predict the flow field around a proposed vehicle. Knowledge of the vehicle flow field at several points along the trajectory enables computation of vehicle surface pressures, aerodynamic forces, and heat transfer characteristics. Mach number and Reynolds number[3] histories are also necessary to: a) characterize the vehicle boundary layer (laminar, transitional or turbulent); b) provide initial conditions for CFD solutions; and c) establish appropriate requirements for ground testing.

In the preliminary design phases the mission requirements establish a vehicle shape based on payload volume and aerothermodynamic limitations for a proposed trajectory. Costs, schedules, launch vehicle availability, reliability and recovery considerations are factored into the design. Development testing and analysis are performed as necessary. As the design progresses more detailed trajectory information becomes available to ensure that stability, guidance, communications, navigation, and

---

[1] At an altitude of 150 km (93.2 mi) a spacecraft or satellite in a circular Earth orbit travels at a speed of 7.8 km/s (25,600 ft/s). In order for a spacecraft to escape the Earth's gravitational field the spacecraft must reach a speed of approximately 11.2 km/s; this is referred to as the Earth escape velocity.

[2] Computational fluid dynamics is the discipline that uses high-speed digital computers to solve the equations of flow around or within various bodies. See section 1.7 for a brief description of computational fluid dynamics.

[3] Mach number is the ratio of the vehicle speed to the speed of sound in the surrounding fluid. It is a measure of the compressibility of the fluid. Reynolds number is the ratio of inertia (or momentum) forces to viscous forces in a fluid flow. It is a function of the speed and size of a body and the viscosity of the fluid. Mach number and Reynolds number are defined in section 1.3.1.

control requirements are satisfied. Validation and qualification tests are performed and the design is baselined. Note that aerothermodynamic design considerations and trajectory calculations are critical in the early and later design phases. Other aspects of spacecraft systems design are discussed in section 3.

1.2   Basic Considerations

Sample return vehicles are designed to ensure that the vehicle payload will survive atmospheric entry. The following conditions must be carefully assessed in the design of an SRV: a) maximum aerodynamic load; b) maximum deceleration load; c) maximum heat flux; and d) total heat load. The maximum aerodynamic load usually occurs when dynamic pressure is a maximum (aerodynamic forces are directly proportional to dynamic pressure). If the vehicle angle of attack[4] is varying, it is necessary to examine the product of dynamic pressure and angle of attack (lift force usually increases as angle of attack increases). The maximum aerodynamic load usually occurs when the product of dynamic pressure and angle of attack is a maximum.

Excessive deceleration loads can be decreased by increasing the vehicle lift-drag ratio L/D. Lifting bodies can be maneuvered to control the SRV impact location. Controlling lift and bank angle[5] can vary both down-range and cross-range impact locations.

Providing an adequate thermal protection system (TPS) is a major design consideration. The structural and heat shield materials selected must be able to withstand the maximum heat flux and the total heat load. Greater heat fluxes are encountered in steep entry than for shallow entry angles.[6] However, the time of flight is usually greater for shallow entry angles, thus increasing the total heat load the vehicle must absorb (since total heat load is equal to the integral of heat flux).[7]

At SRV entry speeds greater than 11 km/s temperatures greater than 11,000 K are encountered in the shock layer. (The region between the bow shock and the vehicle surface is called the shock layer; see fig. 1.) At these temperatures, the heat flux to the vehicle due to radiation from shock-layer gases can be significant. At lower velocities and temperatures the heat flux is primarily convective. Ceramic tiles used on the Space Transportation System (STS) Orbiter and ablating materials[8] used on the Apollo command module and planetary entry vehicles protected these vehicles from high temperatures and high heat loads. Figure 2 shows shock layer temperature as a function of altitude and vehicle velocity for equilibrium air (equilibrium and nonequilibrium flows are discussed in sec. 1.3.2). Note that thermodynamic and transport properties (e.g., viscosity and thermal

---

[4] The angle of attack is the angle between the vehicle axis and the vehicle velocity vector (see fig. 12).

[5] The bank, or roll angle is the angle between a vertical (or horizontal) reference plane and the lift vector.

[6] The flightpath angle is the angle between the horizontal and the vehicle velocity vector; the positive direction is up and the flightpath angle at entry is always negative. (A flightpath angle of –20 deg is shallower than a flightpath angle of –40 deg.)

[7] In this document, the heat flux q is defined as the heat transfer rate per unit area. The symbol q (or q-dot) is often used for heat flux. The heat load Q is the time integral of heat flux. In forced convection heat transfer, the heat flux is proportional to the temperature gradient normal to the flow direction at the body surface. The constant of proportionality k is the transport property known as the thermal conductivity of the fluid.

[8] Ablating materials are materials used on the surface of an entry vehicles to absorb heat by removal of mass, thus blocking the transfer of heat to the vehicle. The total heat load determines the amount of material required to protect the vehicle.

conductivity) of air are not well understood at temperatures above 9000 K. Other features of hypersonic flow around a blunt body are described in the sections that follow.

1.3 Flow Regimes

Atmospheric entry trajectories are shown as functions of altitude and velocity in figure 3 for Space Shuttle Orbiter, Apollo 4, and for a typical SRV (solid lines). Figure 3 also shows approximate boundaries for vibrational excitation, oxygen and nitrogen dissociation, ionization, and thermochemical nonequilibrium (dashed lines). At high altitudes free molecular flow effects are significant (above 150 km for the Orbiter) and continuum flow (see sec. 1.3.1) approximations are not valid. At low altitudes continuum subsonic, transonic, supersonic and hypersonic flows occur. At vehicle speeds below about 100 m/s the flow can be considered incompressible (air density is assumed constant); water is an example of an incompressible fluid. For each flow regime it is necessary to consider the flow field around the vehicle, the vehicle surface pressure distribution, and the heat transfer to or from the vehicle.

1.3.1 Continuum Flow

For air at standard sea level conditions, the distance a molecule travels between collisions is extremely small and the atmosphere appears to be a continuous medium to an object moving through it.[9] Continuum flow occurs when the molecular mean free path λ is much smaller than a characteristic vehicle dimension L; i.e., $\lambda \ll L$. Knudsen number Kn is defined as the ratio of λ to L. Knudsen number can also be expressed as a function of Mach number and Reynolds number. Continuum flow occurs when Kn $\ll$ 1 and free molecular flow occurs when Kn $\gg$ 1. In terms of Mach number M and Reynolds number Re, continuum flow occurs when $M/(Re)^{1/2} < 0.01$ and free molecular flow occurs when M/Re > 3 (Regan and Anandakrishnan, 1993); see table 1. Free molecular and continuum flow regime boundaries are shown as functions of Mach number and Reynolds number in figure 4. Mach number is defined below. Reynolds number is defined in section 1.5.

TABLE 1. FLOW REGIME BOUNDARIES

Flow Regime	Anderson (1989)[a]	Regan and Anandakrishnan (1993)
Free molecular	Kn $\gg$ 1	Re < M/3
Near free molecular	Kn > 1.0	—
Transitional	1.0 > Kn > 0.03	$M/3 < Re < 10{,}000\ M^2$
Continuum	Kn < 0.2	$Re > 10{,}000\ M^2$

[a] Note the overlap of the continuum and transitional flow regimes. Anderson (1989) notes that in this region (0.03 < Kn < 0.2) temperature and velocity slip effects (discontinuities) are present at the body surface.

---

[9] At sea level the molecular mean free path for air is 6.633×10^{-8} m; at 100 km altitude the mean free path is 0.141 m (see National Oceanic and Atmospheric Administration, U.S. Standard Atmosphere, 1976).

A perfect gas is both thermally perfect and calorically perfect. A thermally perfect gas obeys an ideal gas equation of state. A calorically perfect gas has constant values of specific heat, independent of temperature; i.e., specific heat at constant pressure c_p and specific heat at constant volume c_v are constant. A perfect gas is a gas that: satisfies an equation of state; has specific heats independent of temperature; and has internal energy that is a function of temperature alone. A perfect gas is also called an ideal gas.[10] (see Allen, 1965 and Glover, 2006). A nonperfect gas is often referred to as a real gas.

The equation of state for a thermally perfect gas is

$$p = \rho\,R\,T$$

where

p = pressure

ρ = mass density

T = absolute temperature

R = gas constant for a specific gas

$R = \boldsymbol{R}\,/\,\boldsymbol{M}$ = 287 J/kg-K for air

$\boldsymbol{R}$ = universal gas constant = 8314.34 J/(kmol-K)

$\boldsymbol{M}$ = molecular weight = 28.9644 kg/kmol (for air at altitudes ≤85 km).

Note that the perfect gas equation of state given above is not valid for hypersonic flow (Mach numbers above five or six) where the isentropic exponent (or ratio of specific heats) γ for air is no longer constant. Figure 5 shows the variation of γ with shock layer temperature and pressure for air (Gordon and McBride, 1994). Only at temperatures below 800 K (or velocities below about 800 m/s) can air be assumed to be calorically perfect.

In continuum flow Mach number M is a very important dimensionless parameter. Mach number is defined as the ratio of the vehicle (or freestream) speed V to the local speed of sound c:

$$M = V/c$$

where

$c = (\gamma\,R\,T)^{1/2}$ for a thermally perfect gas

γ = isentropic exponent = c_p/c_v

c_p = specific heat at constant pressure

c_v = specific heat at constant volume.

---

[10] The terms perfect fluid and ideal fluid usually refer to an inviscid fluid that may also be incompressible.

An approximation for the speed of sound in thermally perfect air with $\gamma = 1.4$ and $T < 1000$ K is

$$c \approx 20.0 \, (T)^{1/2} \text{ m/s} .$$

Figure 2 shows the effect of freestream, or vehicle, velocity on shock layer temperature. Figure 5 shows the effect of shock layer temperature and pressure on the isentropic exponent γ for air. These variations are important considerations in hypersonic flow.

Mach number is a measure of the effect of the compressibility of a fluid. When Mach number is near or equal to zero the fluid is considered incompressible and the density of the fluid is assumed constant. Flows with Mach number less than one and without shock waves are classified as subsonic. As the Mach number increases, compressibility effects become significant and density can no longer be assumed constant (compressible subsonic flow).

Analytical solutions for incompressible flow around various body shapes can be obtained by solving the Laplace equation for the velocity potential (Karamcheti, 1980). Numerical solutions for subsonic flow around arbitrary two- and three-dimensional bodies can be obtained using panel methods (Katz and Plotkin, 1991).

In incompressible flow static pressure p can be obtained from Bernoulli's equation:

$$p + 1/2 \, \rho \, V^2 = \text{const}$$

where

$$\rho = \text{const for } M \ll 1.$$

For compressible flow ($\rho \neq \text{const}$)

$$\gamma/(\gamma - 1) \, p/\rho + 1/2 \, V^2 = \text{const}$$

It is often useful to determine a pressure coefficient distribution to characterize a flow field. The pressure coefficient is defined as

$$C_p \equiv (p - p_\infty) / q_\infty$$

where q_∞ is the freestream dynamic pressure

$$q_\infty \equiv 1/2 \, \rho_\infty \, V_\infty^2$$

or for a thermally perfect gas

$$q_\infty = 1/2 \, \gamma \, p_\infty \, M_\infty^2 .$$

Note that the subscript ∞ indicates conditions far upstream of the body. The subscript is dropped in later sections when it is not necessary to distinguish between freestream and local conditions.

Similarity rules – Similarity rules can be used to determine lift, drag and pressure coefficients from ground test results or from CFD solutions for similar body shapes. That is, the results for one body shape can be scaled to another slightly thicker or thinner similar body at a slightly lower or higher Mach number.

For compressible subsonic flow (Mach numbers less than about 0.7), pressure and force coefficients on slender bodies at small angles of attack α can be scaled using the Prandtl-Glauert and Goethert similarity rules (Liepmann and Roshko, 1957).

Body 1 has a fineness ratio (body thickness/body length) $\tau = \tau_1$, a freestream Mach number $M = M_1$, and a pressure coefficient $C_p = C_{p1}$; and similarly for Body 2. Assume τ_1 and $\tau_2 \ll 1$ and α_1 and $\alpha_2 \ll 1$ radian.

a) If $\tau_2 / (1 - M_2^2)^{1/2} = \tau_1 / (1 - M_1^2)^{1/2}$, then

$$C_{p2} = C_{p1}$$

If $M_1 \approx 0$, then

$$C_{p2} = C_{p1} = C_{p\,INCOMP}$$

b) If $\tau_2 = \tau_1$, then

$$C_{p2} = C_{p1} (1 - M_1^2)^{1/2} / (1 - M_2^2)^{1/2}$$

If $M_1 = 0$, then $C_{p1} = C_{p\,INCOMP}$ and

$$C_{p2} = C_{p\,INCOMP} / (1 - M_2^2)^{1/2}$$

c) If $M_2 = M_1$, then

$$C_{p2} = C_{p1} \, \tau_2 / \tau_1$$

The Goethert rule applies to slender, axially-symmetric, as well as two-dimensional, bodies:

If $\tau_2 = \tau_1 (1 - M_1^2)^{1/2} / (1 - M_2^2)^{1/2}$ then

$$C_{p2} = C_{p1} (1 - M_1^2)^{1/2} / (1 - M_2^2)^{1/2}$$

There are also transonic, supersonic and hypersonic similarity rules (Van Dyke, 1951 and Spreiter, 1982). The supersonic similarity rule is similar to the subsonic rule with $(1 - M^2)^{1/2}$ replaced by $(M^2 - 1)^{1/2}$.

6

Compressibility and flow regimes – Compressibility is the fluid property that relates changes in pressure and specific volume (or fluid density). Liquids are highly incompressible and gasses are highly compressible. However, at very low speeds gasses behave much like an incompressible fluid. As pressure increases specific volume decreases (and density, the inverse of specific volume, increases).

An incompressible flow is one in which the fluid density remains constant. At low subsonic speeds (M << 1) fluids can be considered incompressible. As vehicle speed increases, changes in the density of the fluid can no longer be neglected and the flow is referred to as compressible (Glover, 2006).

As the speed of a vehicle approaches the speed of sound (M = 1 or "Mach one") shock waves begin to form and the flow becomes transonic. This is the region of the so-called "sound barrier," where a vehicle experiences a large increase in drag and often experiences unsteady buffeting.

Compressibility effects become even more significant at Mach numbers above one. Shock waves are present and distinct and the flow is called supersonic or hypersonic. In supersonic continuum flow the thickness of the shock wave is negligible and flow properties are treated as discontinuities across the shock. Shock waves can be attached (pointed body), detached (blunt body), normal, or oblique (see sec. 1.5). A pointed body can have a detached shock at low supersonic Mach numbers or at any Mach number if the nose semivertex angle is large. A normal shock wave occurs when a supersonic flow is brought to rest at a stagnation point[11] on a body surface normal to the flow direction (or a stagnation line on a wing-like, two-dimensional body). A normal shock wave can also occur in a shock tube when a diaphragm separating high and low pressure gases is removed or ruptured (i.e., as the high pressure gas expands into the low pressure region). Attached oblique shock waves occur when a supersonic flow impinges on a wedge or an inclined plane (if the flow deflection angle is not too large or the Mach number is not too low).

At Mach numbers above around five, the flow is usually considered hypersonic. (The number five is somewhat arbitrary and may depend on the "bluntness" of the body.) Hypersonic flows are characterized by shock waves very close to the body surface (thin shock layer), thick boundary layers, high temperatures, and aerodynamic coefficients that can be nonlinear functions of angle of attack. The "thick" and "thin" characterizations are with respect to a body dimension such as diameter or length. Some features of hypersonic flow over a blunt body are shown in figure 1. At the rearward facing base of the body the flow separates and creates a region of recirculating flow bounded by dividing streamlines. Note that viscous effects are confined to the boundary layer and wake regions.

1.3.2 Hypervelocity Flow

At temperatures above about 800 K at an air pressure of 1 atm, vibrational excitation of O_2 and N_2 molecules becomes significant and specific heats c_p and c_v are no longer constant. The perfect gas equation of state given above continues to apply until dissociation of O_2 begins at about 2500 K. In this temperature range the gas is thermally perfect, but not calorically perfect (Anderson, 1989). At temperatures above about 2500 K the gas is a mixture of chemically reacting gases and each

---

[11] A stagnation point is a point in the flow about a body where the fluid particles have zero velocity with respect to the body.

chemical species obeys the perfect gas equation. For chemically reacting gases it is important to track the concentrations of each species. The velocities producing very high temperatures are often referred to as hypervelocities rather than hypersonic velocities.

Dissociation – The dashed lines on the left side of figure 3 indicate that oxygen and nitrogen molecules in the air behind (or downstream of) a normal shock wave begin to dissociate to form atoms. This is the first significant departure from a calorically perfect gas. For air at a pressure of 1 atm, O_2 dissociation occurs between 2500 K and 4000 K; N_2 dissociation occurs between 4000 K and 9000 K (Anderson, 1989).

Ionization – To the right of the dashed line on the right side of figure 3 ionization effects become important. Electrons leave the oxygen and nitrogen atoms to form ions. This is the cause of the well-known communications "blackout" that occurs during re-entry. For air at a pressure of 1 atm, ionization of O and N molecules occurs at temperatures above 9000 K (Anderson, 1989). Note that NO molecules can become ionized at lower temperatures and pressures.

Nonequilibrium flow – An equilibrium gas flow is one in which the composition of the gas at any point is independent of time and energy is constant along streamlines. A nonequilibrium flow is one that is not in thermal equilibrium or chemical equilibrium; this occurs at very high temperatures and/or in very high-speed flows (large kinetic energies).

Thermal, or thermodynamic, equilibrium means that physical properties (e.g., pressure and temperature) have ceased to change with time at a given point; or that they are varying so slowly that for any point, one can assume thermodynamic equilibrium in some neighborhood about that point. Chemical equilibrium means all chemical reactions are in balance and the system does not undergo any further change in chemical composition or in the concentrations of the various substances.

Frozen flow is a special case of nonequilibrium flow. In frozen flow all chemical reactions are assumed to have stopped—frozen in time. Air in a frozen state can have properties significantly different than equilibrium air at the same temperature and pressure.

The frequency of molecular collisions decreases as altitude increases; and at altitudes above about 50 km, the rate of physical and chemical processes must be considered. A finite time is required for chemical reactions to reach an equilibrium composition. For a brief interval (a few milliseconds) after the temperature increase, the process will be in chemical nonequilibrium. In nonequilibrium flows, specie concentrations and thermodynamic properties are functions of time. Equilibrium properties for chemically reacting gases can be derived from statistical thermodynamics. Computational modeling of nonequilibrium thermodynamics is very difficult and equilibrium thermodynamics is often used as a simple approximation to the actual flow.

Other hypervelocity flow effects – Other effects that must be considered in hypervelocity flow include: a) the need to dissipate enormous quantities of energy (a very large percentage of the vehicle kinetic energy is transferred to the gas in the shock layer); b) very high shock layer temperatures (as high as 14,000 K for sample return missions) are sources of heat radiation that can have a significant effect on vehicle surface temperatures; and c) foreign gases (from vaporization), solid particles (from spallation[12] due to thermal stresses), and products of ablation produce additional complications in the flow field.

1.3.3   Free Molecular Flow

At high altitudes collisions between molecules are less frequent and the significant flow parameter is the Knudsen number Kn. The Knudsen number is the ratio of the molecular mean free path λ and a characteristic vehicle dimension L (usually nose radius or body length):

$$Kn \equiv \lambda/L$$

In general, free molecular flow occurs when Knudsen number is much greater than 1 and continuum flow occurs when Knudsen number is much less than 1 (see table 1). In free molecular flow a distinct bow shock does not exist in front of a blunt body (Rohsenow et al., 1985).

In free molecular flow and at altitudes just below the free molecular limit, molecules near the vehicle surface may have a mean velocity different from that of the surface. This is referred to as velocity slip. If the fluid molecules do not acquire the energy corresponding to the energy of the vehicle surface after one collision, there is lack of momentum accommodation. If a temperature discontinuity occurs at the vehicle surface, there is a lack of thermal accommodation.

In free molecular and near free molecular flows, thermal and momentum accommodation coefficients[13] are can be determined experimentally and used to predict temperature jump and velocity slip at the surface. Thermal accommodation coefficients are functions of gas temperature T, surface temperature T_W, gas composition, and surface material. For air with $T \approx T_W \approx 300$ K, the thermal accommodation coefficient varies from 0.89 to 0.97 for polished metal surfaces (Rohsenow et al., 1985).

The convective heat flux in free molecular flow can be determined from the kinetic theory of gases or obtained from an experimental correlation of Stanton number St and thermal accommodation coefficient. The Stanton number is the ratio of the convective heat transfer at the surface to the heat transferred by the heat capacity of the gas.

$$St \equiv (\text{convective heat transfer coefficient}) / \rho \, c_p \, V$$

where

$$\text{convective heat transfer coefficient} = q / (T_W - T_F)$$

and

$$T_F = \text{characteristic fluid temperature (usually mean or freestream).}$$

---

[12] High thermal stresses at the vehicle surface may cause solid particles to spall and enter the flow field.

[13] The accommodation coefficient is the ratio of the average energy actually transferred between a surface and impinging gas molecules which are scattered by the surface to the average energy which would theoretically be transferred if the impinging molecules reached complete thermal equilibrium with the surface before leaving the surface (Glover, 2006).

Free molecular drag coefficients can be determined from momentum accommodation coefficients. Algorithms for calculating free molecular lift and drag coefficients are given in Regan and Anandakrishnan, 1993. Direct simulation Monte Carlo[14] (DSMC) computational methods are used to determine free molecular and near free molecular flow-field characteristics.

1.3.4   Transition from Free Molecular to Continuum Flow

A vehicle leaves the free molecular flow regime as it descends into the sensible atmosphere. Since Knudsen number is inversely proportional to a vehicle characteristic length, smaller bodies experience free molecular flow at lower altitudes (see table 2). The Space Shuttle Orbiter experiences free molecular flow at altitudes above 170 km and continuum flow at altitudes below around 90 km.

TABLE 2. APPROXIMATE ALTITUDE LIMITS FOR FREE MOLECULAR, TRANSITIONAL, AND CONTINUUM FLOW REGIMES, 1976 U.S. STANDARD ATMOSPHERE*

Characteristic body length, L, m	Free molecular flow regime, altitude, km (Kn > 100)	Transitional flow regime,[a] altitude, km (1 > Kn > 0.03)	Continuum flow regime, altitude, km (Kn < 0.01)
1	>170	110 − 90	<85
0.5	>160	105 − 85	<80
0.1	>130	95 − 75	<70

* National Oceanic and Atmospheric Administration, 1976.
[a] Anderson, 1989.

At altitudes just below the free molecular limit velocity slip and lack of momentum accommodation may occur near the vehicle surface. Near the free molecular flow limit there may also be a jump in temperature at the vehicle surface.

At altitudes just below the free molecular flow limit there is a transitional flow regime. (This is not the same as the transition from laminar flow to turbulent flow in the boundary layer that occurs as Reynolds number increases.) Accurate prediction of aerodynamic coefficients and heat fluxes is difficult near the free molecular flow limit and in the transitional flow regime.

At altitudes just below the transitional flow regime there is a viscous merged layer flow regime. In the transitional flow regime shock waves can have appreciable thickness. (In continuum flow the thickness of a shock wave is negligible and flow properties are treated as discontinuities at the shock location). In the viscous merged layer regime there may be strong interactions between the shock wave and the boundary layer. (Regan, 1984 and Anderson, 1989). Approximate flow regime boundaries are given in tables 1 and 2 and shown in figure 4.

---

[14] The direct simulation Monte Carlo (or DSMC) method is used to simulate gas flows in the free molecular flow regime. It is a computational technique that models the statistical behavior of the gas molecules.

1.4 Shock Waves

Shock waves are a prominent feature of re-entry flows. In continuum flow the thickness of a shock wave is neglected and flow properties are discontinuous at the shock location. A normal shock wave is shown schematically in figure 6a. (A normal shock wave is perpendicular to the upstream velocity vector.) Static pressure p, static density ρ, and static temperature T increase across the shock wave while Mach number M and flow speed V decrease ($p_2 > p_1$, $\rho_2 > \rho_1$, $T_2 > T_1$, $M_2 < M_1$, $V_2 < V_1$) where the subscripts 1 and 2 indicate conditions just upstream and just downstream of the shock, respectively. For a normal shock wave, the downstream Mach number M_2 is always subsonic ($M_2 < 1$). A normal shock wave occurs ahead of a blunt body in supersonic or hypersonic flow and is referred to as a detached or bow shock (see fig. 1). The distance between a detached shock and a body is called the shock-wave standoff distance.

An oblique shock wave occurs when the flow direction changes across the shock and the shock wave is no longer normal to the freestream direction as shown figure 6b. The component of the velocity vector that is normal to an oblique shock wave must be supersonic upstream of the shock and subsonic downstream of the shock. Note that the Mach number downstream of an oblique shock wave may be supersonic; it is only the velocity component normal to the oblique shock wave that is subsonic.[15]

1.5 Boundary Layers and Viscous Effects

Another very important variable in fluid dynamics is the Reynolds number, which is a measure of the effect of fluid viscosity. The nondimensional Reynolds number Re is the ratio of inertia forces to viscous forces; viscous effects become less important as Reynolds number increases. Reynolds number is defined as

$$Re \equiv \rho \, V \, L \, / \, \mu$$

where

ρ = mass density of fluid

V = speed of flow (or vehicle speed)

L = characteristic vehicle dimension

μ = absolute viscosity of fluid = $\mu(T)$.

The Sutherland approximation for the absolute viscosity of air is given in section 2.

---

[15] Equations, tables and charts for flow properties across normal and oblique shock waves are given in NACA Report 1135 (Ames Aeronautical Laboratory, 1953). Scanned images of NACA reports are available on the NASA Technical Reports Server (see References). A website is available to calculate conditions across normal and oblique shock waves and other compressible flow properties (Delft Technical University, 2006).

Reynolds number can also be expressed in terms of kinematic viscosity ν

$$Re = V L / \nu$$

where

$$\nu = \mu / \rho.$$

In inviscid (nonviscous) flows, viscosity effects are neglected and the fluid velocity at the surface of a body can have a finite, nonzero value. In reality, fluid viscosity requires the velocity to be equal to zero at the body surface (no-slip condition). At moderately high Reynolds numbers (about 100,000) the effect of viscosity is appreciable only in the vicinity of the body. This discovery by the German scientist Ludwig Prandtl was one of the most significant in aeronautics and is the basis for boundary layer theory. The development of a boundary layer in the flow over a flat plate is shown schematically in figure 7.

In figure 7 the flow along the plate starts out laminar at the leading-edge stagnation point where the flow velocity is equal to zero. As the local Reynolds number Re_x based on the distance x from the leading edge increases, the flow ceases to be laminar and passes through a "transition zone" before becoming fully turbulent.[16] Boundary layer transition is affected by several factors including the roughness of the body surface and the turbulence in the freestream. Heat transfer and skin friction (or viscous) drag are strongly dependent of whether the boundary layer is laminar or turbulent. The boundary layer thickness is defined as the distance where the local longitudinal velocity $V(y)$, the velocity parallel to the surface of the plate, becomes equal to 99 percent of the freestream velocity V.

For a flat plate in incompressible flow, the laminar boundary layer thickness δ_{LAM} can be expressed as a function of local Reynolds number and distance from the leading edge of the plate x

$$\delta_{LAM} = 5.0 \, x / Re_x^{1/2}$$

where

$$Re_x = \rho \, V \, x / \mu$$

and ρ and V are freestream density and freestream velocity and μ is the absolute viscosity of fluid.

The skin-friction drag coefficient due to shear stress C_f is equal to τ_W / q , where τ_W is the shear stress at the body surface and q is the freestream dynamic pressure. In laminar, incompressible flow over a flat plate of length L the skin-friction drag coefficient is obtained from the Blasius drag equation (see Schlichting, 2000):

$$C_{f\,LAM} = 1.328 / Re_L^{1/2}$$

---

[16] A laminar flow is one in which a fluid flows in parallel layers. In turbulent flow, the fluid velocity vector has unsteady, random components added to its mean value. Examples of transition from laminar to turbulent flow are cigarette smoke rising in still air and a stream of ink flowing in clear water. In both cases the flow starts out as a uniform, laminar flow. As the distance from the point of origin of the flow increases (or as the flow velocity increases) the flow will abruptly begin to pass through a transition regime and the smoke or ink will begin to mix, diffuse, and become turbulent.

where

$$Re_L = \rho \, V \, L \, / \, \mu.$$

For a flat plate in incompressible flow, the turbulent boundary layer thickness δ_{TURB} can also be expressed as a function of Re_x and x

$$\delta_{TURB} = 0.37 \, x \, / \, Re_x^{1/5}$$

The skin-friction drag coefficient for a flat plate of length L in turbulent flow is

$$C_{f \, TURB} = 0.074 \, / \, Re_L^{1/5}$$

If the transition from laminar to turbulent flow is assumed to occur at $Re_T \approx 500{,}000$, then the skin friction drag coefficient can be estimated from the Prandtl-Schlichting formula for a smooth flat plate at zero angle of attack:

$$C_{f \, TRANS} = 0.455 \, / \, (\log Re_L)^{2.58} - 1700 \, / \, Re_L$$

$$\text{for } 5 \times 10^5 < Re_L < 10^8$$

These relations are shown in figure 8 (see Schlichting, 2000 and White, 2005). The skin friction drag force can be determined from the equation:

$$\text{skin friction drag force} = C_f \, q \, S_W$$

where S_W is the wetted surface area.

It is very difficult to predict the location and extent of the transition region. Usually a transition line location is assumed based on an empirical correlation from experimental data for similar body shapes and conditions. The correlation is usually between Reynolds number based on distance from the leading edge of a wing or the stagnation point on a blunt body, and the Mach number at the edge of the boundary layer. Upstream of the transition line the flow is assumed laminar and downstream the flow is assumed turbulent. The location of the transition line is very important for accurate prediction of skin friction drag and aerodynamic heating. Anderson (1989) notes that uncertainties in the transition region location can have a very significant effect on vehicle weight estimates. In addition to surface roughness and freestream turbulence, other factors that can influence boundary layer transition are Mach number, Reynolds number, body shape (blunt versus slender) and nose radius, wall and stagnation temperatures, ablation (mass removal at the surface), angle of attack, axial pressure gradient (decreasing pressure promotes transition), and chemical reactions in the flow (Tauber, 1990).

The variation of skin friction coefficient and Stanton number with Reynolds number is similar to that shown in figure 8 for skin friction drag coefficient. There are a variety of methods for predicting skin friction in high-speed flow (see Liepmann and Roshko, 1957; Anderson, 1989; and White, 2005). For a given Reynolds number, the skin friction coefficient generally decreases as Mach number increases.

1.6 Flight and Ground Testing

Flight and ground testing are used to determine vehicle aerodynamic and aerothermodynamic characteristics. Flight tests to simulate SRV atmospheric entry conditions are relatively expensive and difficult to perform (see sec. 7). One possibility is to launch a model vehicle into the atmosphere with a propulsive device. Ground testing is more often employed to simulate SRV flight conditions. Many different types of ground test facilities have been used to obtain vehicle pressure distributions and aerodynamic force and moment[17] characteristics. Heat transfer measurements and flow visualization are possible in conventional ground test facilities. In a ballistic range, data are obtained from a model launched from a gun into a quiescent gas (Chapman and Yates, 1998). A shock tube can be used to achieve higher stagnation temperatures than a conventional wind tunnel. However, shock tube test durations are extremely short and data acquisition is much more difficult. Arc-heated (or arc jet) test facilities such as those at the NASA Ames Research Center and U.S. Air Force's Arnold Engineering Development Center (AEDC) in Tullahoma, Tennessee are usually used to simulate high-temperature effects.

The principal flight variables to match for atmospheric entry simulation are:

Mach number (compressibility effects)

Reynolds number (viscous effects)

Total enthalpy (thermal energy)

Ratio of specific heats (isentropic exponent of appropriate test gas)

Stagnation pressure (or flight altitude)

Ratio of wall temperature to stagnation (or freestream) temperature

Viscous interaction parameter[18] (for some hypervelocity flows)

Gas chemistry (including dissociation, ionization, and ablation effects)

Knudsen number (for free molecular or rarefied atmosphere effects)

Prandtl number[19] (heat conduction)

It is impossible to achieve complete simulation of entry flight conditions in a single test facility. Usually an attempt is made to match the most important variables in one or more facilities. In conventional wind-tunnel and ballistic-range testing, flight Mach number and Reynolds number can often be matched. Low-density wind tunnels are used to simulate free molecular flows by matching the flight Knudsen number. For hypervelocity flows, arc-jet facilities are used to evaluate TPS materials. In arc-jet testing an attempt is made to match total enthalpy and stagnation pressure. Note that arc-jet tests can not provide useful aerodynamic characteristics.

---

[17] In aerodynamics as in rigid-body mechanics the moment of an aerodynamic force is the tendency of the force to rotate a body about a point or an axis. The moment is the product of the force and a moment arm defined as the perpendicular distance from the point or the axis to the line of action of the force. Moments are vector quantities having both magnitude and direction.

[18] Viscous interaction parameter is defined in section 3.

[19] The Prandtl number is the ratio of momentum transfer to heat transfer. Prandtl number is defined in section 3.

1.7 Computational Fluid Dynamics and the Navier-Stokes Equations

The Navier-Stokes equations describe the motion of a viscous fluid. Changes in momentum, or acceleration, of fluid elements are related to changes in pressure and viscous forces to determine the motion and properties of the fluid. The equations can be derived from the basic principles of conservation of mass, momentum, and energy. The Navier-Stokes equations are extremely useful in the study of fluid dynamic flows; they are higher-order partial differential equations and are very difficult to solve. There are limitations and assumptions involved in the derivation; e.g., the fluid must be continuous and homogeneous (it does not contain voids or bubbles). In order to obtain useful solutions to practical problems it is necessary to use computational fluid dynamics (CFD) techniques. For an inviscid (nonviscous) fluid the Navier-Stokes equations reduce to the Euler equations.[20]

Computational fluid dynamics methods are used to obtain detailed flow field characteristics for SRVs. The calculated surface pressures and temperatures are used to determine aerodynamic forces and heat loads. However, it is absolutely necessary to validate and calibrate CFD methods and solutions. The necessary validation and calibration can be accomplished by means of flight or ground test.

Computational methods have been used in fluid dynamics for several decades. Early efforts were for inviscid, low-speed and supersonic flows over simple bodies. In the 1980s high-speed computers (and a greater understanding of the stability of numerical methods) made possible the solution of viscous flows around complicated body shapes (CFD-Online, 2006).

In CFD the space around the body is divided into small cells to form a mesh or grid and super computers are used to iteratively solve the compressible or incompressible, laminar or turbulent fluid flow equations. The equations for chemical reactions, species concentration, and heat transfer can also be included (Anderson, 1995). CFD solutions are very sensitive to the mesh configuration and turbulence model selected.

Note that there are also numerous computer codes available for heat transfer modeling, structural design, trajectory analysis, chemistry and thermodynamic properties. Many of these codes were originally developed by NASA and its U.S. space program contractors (Open Channel Software, 2006).[21]

---

[20] NASA Glenn Research Center has a Learning Technology website that describes the Navier-Stokes equations and includes many other subjects in aeronautics (see NASA Glenn Research Center).

[21] The Open Channel Foundation has an agreement with the National Technology Transfer Center to maintain NASA's COSMIC software collection. This collection covers a wide range of disciplines including engineering, chemistry, and aerodynamics.

2. EARTH ATMOSPHERE AND ACCELERATION DUE TO GRAVITY

2.1 Properties of Earth's Atmosphere

Knowledge of the properties of Earth's atmosphere is necessary to determine aerodynamic and thermal loads during entry. Pressure, temperature, mass density, species concentration, molecular weight, viscosity, thermal conductivity and, at higher altitudes, mean free path and collision frequency must be known as functions of altitude. Below an altitude of 100 km Earth's atmosphere is primarily nitrogen and oxygen molecules (O_2 dissociation becomes significant at altitudes above 100 km); see table 3 for 1976 U.S. Standard Atmosphere values (National Oceanic and Atmospheric Administration, 1976).

TABLE 3. PRIMARY CONSTITUENTS OF EARTH'S ATMOSPHERE IN MOLE (VOLUME) FRACTIONS, 1976 U.S. STANDARD ATMOSPHERE*

Altitude, y, km	N_2	O_2	O	Ar
0 to 85	0.7808	0.2095	0	0.0093
100	0.7750	0.1798	0.0354	0.0079
125	0.7085	0.0777	0.2111	0.0022
150	0.6024	0.0530	0.3430	0.0010

* National Oceanic and Atmospheric Administration, 1976.

Properties of the atmosphere vary with time of day, day of the year and solar radiation intensity. However, these distinctions are usually ignored for hypervelocity Earth entry (Regan, 1984) and a standard atmosphere is adopted for trajectory simulation and load calculation. Above an altitude of 100 km, atmospheric density (and drag force) is so low that its influence on entry vehicle dynamics is usually considered negligible.

Properties of the Earth's atmosphere at sea level are given in table 4 (National Oceanic and Atmospheric Administration, 1976). Properties for the 1976 U.S. Standard Atmosphere can be obtained for altitudes up to 150 km from a convenient and easy to use Internet database (Squire, 2006).

TABLE 4. SEA LEVEL PROPERTIES OF EARTH'S ATMOSPHERE, 1976 U.S. STANDARD ATMOSPHERE*

Quantity	Metric Units	English Units
Atmospheric density, ρ_0	1.2250 kg/m^3	0.0765 lb/ft^3
Atmospheric pressure, p_0	101,325 Pa	2116 lb/ft^2
Atmospheric temperature, T_0	288.15 K	518.67 °R
Mean free path, λ_0	6.6328 x 10^{-8} m	21.76 x 10^{-8} ft
Molecular weight, $\boldsymbol{M_0}$	28.9644 kg/kmol	—
Speed of sound, c_0	340.294 m/s	1116 ft/s
Thermal conductivity, k_0	0.025326 W/(m-K)	0.0146 Btu-ft/(hr-ft^2-°R)
Absolute viscosity, μ_0	1.789 x 10^{-5} kg/(m-s)	1.202 x 10^{-5} lb/(ft-s)
Acceleration due to gravity, g_0	9.80665 m/s^2	32.174 ft/s^2

* National Oceanic and Atmospheric Administration, 1976.

For Earth's atmosphere the mass density of air ρ and acceleration due to gravity g can be expressed as functions of altitude (see secs. 2.1.2 and 2.4).

Static pressure p can be determined by summing the pressure and gravitational forces acting on a fluid element:

$$dp/dy = - \rho\, g$$

$$p = p_0 - \int \rho\, g\, dy$$

where geometric altitude y is measured from the mean Earth radius R_0 (see sec. 2.4).

Note that geopotential altitude h is defined as follows:

$$g_0\, dh = g\, dy$$

or

$$h = \int [g(y) / g_0]\, dy.$$

2.1.1 Temperature

The 1976 Earth model atmosphere consists of constant temperature layers and layers with temperature varying linearly with altitude. Between sea level and about 100 km there are four distinct layers (see fig. 9).

Troposphere (below 11 km) the temperature gradient (or lapse rate) is –6.5 K/km.

Stratosphere (from 11 km to about 47 km) the temperature gradient varies from 0 to 2.8 K/km.

Mesosphere (from about 47 km to 80 km) the temperature gradient varies from 0 to –3.9 K/km.

Thermosphere (above 80 km) the temperature gradient varies from 0 to 9 K/km.

The bounding altitudes separating the four regions are referred to as: the Tropopause (11 km); the Stratopause (about 47 km); and the Mesopause (80 km).

2.1.2   Density

This simple expression is often used to determine the atmospheric mass density ρ as a function of altitude y

$$\rho(y) = K\, e^{-\beta\, y}$$

where K and β are constants selected to match the preferred model atmosphere; β is called the inverse density scale height.

Allen and Eggers (1958) and Eggers et al. (1958), used values of 0.0034 slugs/ft^3 and 22,000 ft for K and $1/\beta$, respectively. Tauber (1998) recommends K = ρ_0 = 1.225 kg/m^3 and $1/\beta$ = 7200 m. Regan and Anandakrishnan (1993) recommend Allen and Eggers' values equal to 1.752 kg/ m^3 and 6700 m for K and $1/\beta$, respectively. Figure 10 shows a comparison between the analytic expression given above for $\rho(y)$ and the 1976 U.S. standard atmosphere.

2.1.3   Mean Free Path

The mean free path λ is the mean value of the distance traveled by a neutral particle between successive collisions with other particles (National Oceanic and Atmospheric Administration, 1976). Mean free path is shown as a function of altitude in figure 11. The limits of the continuum and free molecular flow regimes are determined by the ratio of the mean free path to a characteristic body length. This ratio is called the Knudsen number.

2.2   Viscosity

The transport property known as absolute, or dynamic, viscosity μ is important in the determination of Reynolds numbers and frictional forces. For air at altitudes less than 85 km, absolute viscosity μ can be determined from the air temperature T by the Sutherland approximation

$$\mu = 1.458 \times 10^{-6}\, T^{3/2} / (T + 110.4) \text{ kg/m-s}$$

where T is in Kelvin.

2.3 Thermal Conductivity

The coefficient of thermal conductivity k is an important transport property in heat transfer. For air at altitudes less than 85 km, thermal conductivity k can be determined from the air temperature T by a formula similar to the Sutherland approximation for viscosity

$$k = 2.646 \times 10^{-3} \, T^{3/2} / (T + 245.4 \times 10^{-12/T}) \text{ W/m-K}$$

where T is in Kelvin (Regan and Anandakrishnan, 1993).

2.4 Acceleration Due to Gravity

The acceleration due to gravity g can be expressed as a function of altitude y

$$g(y) / g_0 = [R_0 / (R_0 + y)]^2 = (1 + y / R_0)^{-2}$$

$$\approx [1 - 2 \, (y / R_0) + 3 \, (y / R_0)^2 + \dots]$$

where

$$g_0 = g(0) = 9.80665 \text{ m/s}^2$$

and

$$R_0 = \text{mean or effective radius of the Earth} = 6{,}356{,}766 \text{ m.}$$

The above approximation for $g(y) / g_0$ is derived from the binomial theorem for $y / R_0 \ll 1$ (see Zwillinger, 2002). Note that the Earth's mean equatorial radius is 6,378,140 m (approximately 21 km larger than its polar radius). The Earth's equatorial bulge and its nonuniform mass density result in deviations from the inverse-square approximation for spherical bodies. These deviations are usually accounted for in trajectory simulation codes. The Earth's oblateness is accounted for by including the 2nd zonal harmonic term J_2 in the gravitational model.

The International Gravity Formula gives g at sea level as a function of latitude λ

$$g_0 = 9.780495 \, [1 + 0.0052892 \, \sin^2(\lambda) - 0.0000073 \, \sin^2(2 \, \lambda)]$$

For $\lambda = 45$ deg this formula gives: $g_0 = 9.806289$ m/s^2 (Glover, 2006).

3. AERODYNAMICS AND HEAT TRANSFER

3.1 Basic Aerodynamic Concepts

The importance of dimensionless parameters, such as Mach number, Reynolds number and Knudsen number, in the study of atmospheric entry was mentioned in section 1. Dimensionless force and moment coefficients are equally important in trajectory computation. If a vehicle's surface pressure distribution is known, the aerodynamic forces and moments can be determined by integrating the pressure over the surface area. Often it is possible to test a scale model of a vehicle in a wind tunnel and measure the forces and moments on the model. If the test Mach number and the test Reynolds number match those in flight, then the forces and moments can be scaled to flight conditions.[22] (If free molecular flow will be encountered in flight, then ground-test Knudsen number should match the flight Knudsen number.)

An aerodynamic force coefficient is defined as

$$\text{Force coefficient} = \text{Force} / (\text{dynamic pressure x reference area})$$

A moment coefficient is defined as

$$\text{Moment coefficient} = \text{Moment} / (\text{dynamic pressure x reference area x reference length})$$

where

$$\text{dynamic pressure} = q = 1/2\ \rho\ V^2$$

or for a thermally perfect gas (not valid for M >> 1)

$$q = 1/2\ \gamma\ p\ M^2$$

ρ = freestream mass density of fluid

V = freestream fluid velocity

M = freestream Mach number

P = freestream static pressure

γ = isentropic exponent.

---

[22] Force Coefficient = $\text{Force}_{\text{FLIGHT}}$ / (q_{FLIGHT} x S_{FLIGHT}) = $\text{Force}_{\text{WIND TUNNEL}}$ / ($q_{\text{WIND TUNNEL}}$ x $S_{\text{WIND TUNNEL}}$)
   or $\text{Force}_{\text{FLIGHT}}$ = $\text{Force}_{\text{WIND TUNNEL}}$ (q_{FLIGHT} / $q_{\text{WIND TUNNEL}}$) (S_{FLIGHT} / $S_{\text{WIND TUNNEL}}$) = Force Coefficient x q_{FLIGHT} x S_{FLIGHT}.

In a wind tunnel test a calibrated strain gauge balance is used to measure forces and moments. The axial force, normal force, and pitching moment are of principal concern:

$$\text{Axial force} = A = C_A \, q \, S$$

$$\text{Normal force} = N = C_N \, q \, S$$

$$\text{Pitching moment} = M = C_m \, q \, S \, L$$

or

$$\text{Axial force coefficient} = C_A = A \, / \, q \, S$$

$$\text{Normal force coefficient} = C_N = N \, / \, q \, S$$

$$\text{Pitching moment coefficient} = C_m = M \, / \, q \, S \, L$$

where

$$S = \text{reference area (cross-sectional or planform area)}$$

$$L = \text{reference length.}$$

Once C_A and C_N are known it is a simple matter to determine the lift and drag coefficients and the lift and drag forces

$$C_L = C_N \cos \alpha - C_A \sin \alpha$$

$$C_D = C_N \sin \alpha + C_A \cos \alpha$$

where α = angle of attack (the angle between the vehicle axis and the velocity vector)

$$\text{Lift force} = C_L \, q \, S$$

$$\text{Drag force} = C_D \, q \, S.$$

Figure 12 shows how axial and normal forces are resolved into lift and drag by vector addition. Note that the vector sum $\mathbf{R} = \mathbf{L} + \mathbf{D} = \mathbf{N} + \mathbf{A}.$

For some body shapes, analytical and/or experimental lift, drag and moment coefficient data are readily available for the entire Mach number range from subsonic to hypersonic. These data can be found in handbooks and textbooks; see Hoerner (1965 and 1985), McCormick (1995), and Avallone and Baumeister (1996).

3.2 Aerodynamic Drag

The aerodynamic force on a vehicle acting directly opposite to the vehicle's velocity vector is called drag. In the absence of thrust, atmospheric drag results in vehicle deceleration. Various components of drag force are described below. Drag estimation can be a very involved process.

Pressure or form drag (usually called wave drag in supersonic or hypersonic flow) is the result of pressure forces acting on the vehicle surface. In hypersonic flow, the Newtonian approximation (described below) usually provides an adequate estimate of the pressure drag.

Skin friction drag is the result of viscous shear stresses acting on the vehicle surface. The skin friction drag coefficient C_f is based on the vehicle surface (or "wetted") area S_W and not on a cross-sectional area.[23] The skin friction drag coefficient increases when the flow in the boundary layer changes from laminar to turbulent. Modern fighter aircraft have skin friction drag coefficients around 0.003 (McCormick, 1995). The skin friction drag coefficient for a flat plate is shown as a function of Reynolds number in Figure 8. Note that at $Re_X = 1,000,000$, the incompressible turbulent skin friction drag coefficient is 3 times greater than the incompressible laminar skin friction drag coefficient.[24]

Base drag is the drag produced by the pressure acting on the base of the vehicle and can be negative (acting in the upstream direction). Base drag is a function of the body and afterbody shapes, Mach number, and Reynolds number.[25]

Another way to look at drag is to consider it to be composed of a component independent of angle of attack (zero-lift drag) and a component that is a function of angle of attack (called induced drag). For wing sections, induced drag usually increases as the square of the lift.

In subsonic flow, drag is relatively unaffected by Mach number. As Mach number approaches one, shock waves begin to form and the drag coefficient begins to increase significantly (transonic flow regime). In supersonic flow, the drag coefficient may continue to increase, remain constant, or decrease depending on the body shape (McCormick, 1995). Typical values of drag coefficient are shown in table 5.

The drag on two-dimensional bodies (e.g., infinite cylinders with axis normal to the freestream direction and wedges of infinite width) is usually greater than the drag of corresponding three-dimensional bodies (spheres and cones).

---

[23] When adding skin friction and pressure drag coefficients the skin friction coefficient must first be multiplied by the ratio of the surface area to the cross-sectional area S_W/S.

[24] Convective heat transfer varies directly as skin friction and follows a similar trend as Reynolds number is increased.

[25] The drag force measured in a wind tunnel may not include base drag if the wind-tunnel model support (usually called a "sting") is attached at the base of the model.

TABLE 5. TYPICAL VALUES OF DRAG COEFFICIENT*

Body Shape	Incompressible ρ = constant $M \approx 0$	Hypersonic $M \gg 1$
Flat Plate		
Surface parallel to velocity vector, $\alpha = 0$ (skin friction drag: see fig. 8)	$1.33/(Re_L)^{1/2}$ (laminar flow)	—
	$0.074/(Re_L)^{1/5}$ (turbulent flow)	—
Surface perpendicular to velocity vector, $\alpha = 90$ deg	1.2 (for $L \approx$ width)	1.8 – 2.0
Circular Disc	1.2	1.8 – 2.0
Circular Cylinder (axis perpendicular to velocity vector) $\quad 1000 < Re_D < 100,000$	0.6 (for $L \approx$ diameter) 1.2 (for $L \gg$ diameter)	1.3 —
Sphere		
$\quad 1000 < Re_D < 300,000$	0.4	1.0
$\quad Re_D \approx 500,000$	0.2	—
Cone		
Semivertex angle $\theta_C = 15$ deg	0.34	0.134[a]
Semivertex angle $\theta_C = 30$ deg	0.51	0.500[a]

*Avallone and Baumeister, 1996; McCormick, 1995; Hughes and Brighton, 1999; and White, 2005.
[a] Inviscid hypersonic flow (does not include skin friction drag), perfect gas, $\gamma = 1.4$ (see table 6).

3.3 Newtonian Impact Theory for Inviscid Hypersonic Flow

A law proposed by Isaac Newton more than three centuries ago can be used to predict lift and drag in inviscid (nonviscous) hypersonic flow (Anderson, 1989). Newton's sine-squared law states that the force acting on an inclined body surface is equal to $\rho V^2 S \sin^2 \theta$

$$F = \rho V^2 S \sin^2 \theta = \Delta p \, S$$

where

θ = angle between tangent to body surface and freestream velocity vector

Δp = surface pressure – freestream static pressure

S = surface area

or expressed in terms of a pressure coefficient C_p

$$C_p \equiv \Delta p / q = 2 \sin^2 \theta$$

where

$$q = 1/2\, \rho\, V^2.$$

This result is of no value in subsonic or supersonic flows and is not applicable to downstream facing surfaces (where $\theta < 0$ and C_p is assumed equal to zero). Inviscid hypersonic aerodynamic force coefficients can be obtained by integrating the Newtonian pressure coefficient over the body surface.

Newtonian lift and drag coefficients C_L and C_D for a flat plate at angle of attack α are

$$C_L = 2 \sin^2 \alpha \cos \alpha$$

$$C_D = 2 \sin^3 \alpha$$

$$L/D = \cot \alpha$$

These quantities are shown as functions of angle of attack in figure 13. Note that the lift coefficient has a maximum value of 0.770 at $\alpha = 54.7$ deg. The value of L/D increases as α decreases. However, L/D will go to zero as α and L go to zero because D has a finite value at $\alpha = 0$ due to skin friction drag.

In inviscid hypersonic flow the Newtonian approximation gives the following values for the wave or pressure drag coefficient:

For a circular cylinder with its axis perpendicular to the freestream direction
$$C_D = 4/3$$

For a sphere
$$C_D = 1$$

For a cone with semivertex angle θ_C
$$C_D = 2 \sin^2 \theta_C$$

A modification to Newtonian theory proposed by Lester Lees gives more accurate results for blunt bodies. Lees' approximation replaces the coefficient 2 in the sine-squared law with the pressure coefficient at the stagnation point behind the normal shock wave $C_{p\,MAX}$ (see Anderson, 1989)

$$C_p = C_{p\,MAX} \sin^2 \theta$$

For a perfect gas, $C_{p\,MAX}$ can be expressed as a function of freestream Mach number M and the isentropic exponent γ (Ames Research Staff, 1953).

As M goes to ∞ $C_{p\,MAX}$ goes to 1.839 (for $\gamma = 1.4$)

25

The following equation gives the Newtonian drag coefficient C_D for a sphere-cone at zero angle of attack with nose radius r_N, base radius r_B, and cone-section semivertex angle θ_C :

$$C_D = 2 \sin^2 \theta_C + (r_N / r_B)^2 \cos^4 \theta_C$$

C_D is shown as a function of bluntness ratio r_N / r_B and cone-section semivertex angle θ_C in figure 14. (When lift is zero, the drag coefficient is referred to as the zero-lift drag coefficient; e.g., axially-symmetric bodies have zero lift at zero angle of attack).

The lift and drag coefficients presented above are for forces due to surface pressure in inviscid hypersonic flow. In supersonic and hypersonic flow the pressure drag is called wave drag because the surface pressure increase is due to flow compression as the flow passes through the shock wave.

Hypersonic lift, drag and pressure coefficients for many body shapes can be readily obtained from simple computer algorithms (Regan and Anandakrishnan, 1993). Note that in this section viscous effects are neglected and only inviscid pressure forces are considered.

3.4 Other Approximate Methods for Inviscid Hypersonic Flow

The static pressure at any point on the surface of a body in hypersonic flow can often be determined from the slope of the body surface at that point. These approximations are referred to as local surface inclination methods (see Anderson, 1989).

3.4.1 Tangent-Wedge Method

The tangent-wedge method can be applied to sharp-nosed, two-dimensional bodies with attached shock waves. The local static pressure at a point on the body surface is assumed equal to the static pressure behind an oblique shock wave with deflection angle equal to the slope of the body surface at the point in question; i.e., the equivalent wedge has deflection angle equal to the local body slope. The local static pressure is a function of freestream Mach number M, the local deflection angle θ and the isentropic exponent γ. Lift and drag can be obtained by integrating the pressure over the body surface. For SRVs two-dimensional body sections are likely to occur only as appendages to the basic body (e.g., wings, fins, external panels, etc.). Flow properties downstream of oblique shock waves are given in NACA Report 1135 (Ames Research Staff, 1953); an Internet calculator is also available (Devenport, 2006).

3.4.2 Tangent-Cone Method

The tangent-cone method for sharp-nosed, three-dimensional (or axially-symmetric) bodies is similar to the tangent-wedge method for two-dimensional bodies. The tangent-cone method also applies only to sharp-nosed bodies with attached shock waves. Static pressure on the body surface is obtained from supersonic conical flow solutions (see Sims, 1964). This reference tabulates the flow properties between the conical shock wave and the cone surface. The tabulated values were obtained from numerical solutions of ordinary differential equations for conical flow with the ratio of specific heats, freestream Mach number, and cone semivertex angle as parameters. Both tangent-cone and tangent-wedge methods give reasonably good results for lift, drag and pressure on sharp-nosed bodies.

26

3.4.3　Shock-Expansion Method

For a body with an attached, or a detached shock wave the flow is first compressed as it passes through the shock wave and it then expands as it flows along the body surface (assuming convex body curvature). A uniform, two-dimensional, supersonic flow over a convex curve is known as a Prandtl-Meyer expansion fan. For a Prandtl-Meyer expansion, the local Mach number is a function of the initial Mach number M, the turning or deflection angle and the isentropic exponent γ. The local static pressure and pressure coefficient can be determined from the local Mach number and the isentropic flow relations; see NACA Report 1135. Lift and drag are obtained by integrating the pressure over the body surface. Although the Prandtl-Meyer expansion is intrinsically two-dimensional it provides reasonably good results for axially-symmetric bodies in hypersonic flow. An Internet calculator is available for obtaining Prandtl-Meyer expansion flow properties (Delft Technical University, 2006).

3.5　Shock Wave Detachment

Sharp-nosed bodies usually have attached shock waves in supersonic and hypersonic flow (oblique shock waves for two-dimensional wedge-shaped bodies and conical shock waves for conical bodies). If the freestream Mach number is decreased or if the apex angle of a pointed body is increased the shock wave will detach and the flow will be similar to that for the blunt body shown in figure 1. Figure 15 shows sharp-nosed bodies with attached and detached shock waves. (Blunt bodies have detached shock waves in supersonic and hypersonic flow.) Minimum Mach numbers and maximum semivertex angles for attached shock waves for cones and wedges in inviscid, perfect gas flow are given in NACA Report 1135. The conical flow charts in NACA Report 1135 are for γ = 1.405 (Kopal, 1947). Conical flow solutions for γ = 1.4 are given in NASA SP-3004 (Sims, 1964). NASA SP-3007 contains solutions for cones at small angles of attack (Sims, 1964).

3.6　Hypersonic Mach Number Independence

Analytical solutions and experimental results show that for very high freestream Mach numbers, many flow characteristics (including lift, drag, pressure and moment coefficients) are relatively insensitive to changes in Mach number. This phenomenon is observed in experimental data at hypersonic Mach numbers for both sharp and blunt, two- and three-dimensional bodies. For sharp circular cones in inviscid, perfect gas flow, table 6 shows that the zero-lift drag coefficient approaches the Newtonian values as the Mach number increases and the change in drag coefficient is very small for freestream Mach numbers greater than ten.

TABLE 6. ZERO-LIFT DRAG COEFFICIENT C_{D_0} FOR SHARP, CIRCULAR CONES, INVISCID, PERFECT GAS WITH $\gamma = 1.4$*

Freestream Mach number, M	Cone semivertex angle, θ_C, degrees		
	10	20	30
1.5	0.124	0.387	—[b]
2.5	0.094	0.299	0.598
3.5	0.083	0.275	0.561
5	0.075	0.260	0.542
7	0.070	0.253	0.533
10	0.067	0.249	0.529
15	0.065	0.247	0.526
20	0.064	0.246	0.525
Newtonian[a]	0.060	0.234	0.500

* NASA SP-3004 (Sims, 1964).
[a] The Newtonian zero-lift drag coefficient is equal to $2 \sin^2 \theta_C$. (For right circular cones at zero angle of attack the zero-lift drag coefficient is equal to the pressure coefficient.)
[b] Shock wave is detached.

3.7 Hypervelocity and High Temperature Effects

3.7.1 Wall Catalysis

In chemically reacting flows over solid surfaces some surface materials promote recombination of atoms. Surfaces that promote complete recombination of all atoms are called catalytic. (If no recombination occurs at the surface, the surface is referred to as noncatalytic.) Wall catalysis is a concern because recombination reactions can increase the local heat flux by a factor of 2 or more. It is also possible to have partially catalytic surfaces that are very difficult to model because of the complexity of nonequilibrium chemistry. Generally a vehicle designer will attempt to select a noncatalytic surface (a surface which does not promote recombination reactions).

Lightweight ceramic tiles with reaction cured glass (RCG) coatings are characterized as noncatalytic TPS materials. Other coatings, such as toughened uni-piece fibrous insulation (TUFI) have been developed to provide low catalycity and increased strength. Metallic surfaces, on the other hand, are highly catalytic and promote recombination reactions. Metallic TPS materials have been proposed (e.g., for reusable launch vehicles) but suffer from the problem of high catalycity as well as high thermal expansion (this makes noncatalytic surface attachment or coating difficult). Partially catalytic TPS materials include reinforced carbon carbon (RCC) and silicon carbide materials.

When using CFD codes, surface catalysis is an important boundary condition and the mass fraction of each chemical species is a dependent variable. For a fully catalytic wall, recombination reactions occur at an infinite rate.

3.7.2 Viscous Interaction

In classical boundary layer theory, it is assumed that viscous effects are confined to a small region adjacent to the surface of the body. In hypersonic flow, boundary layer thickness increases as the square of the Mach number. Thus for high Mach number flows, the classical small boundary layer thickness assumption ($\delta \ll L$) is no longer valid and there is significant interaction between the viscous (boundary layer) and the nonviscous regions of the flow field. This is referred to as the viscous merged layer flow regime and it occurs at higher altitudes and velocities than continuum flow.

Hypersonic vehicle surface pressures can be correlated using a viscous interaction parameter χ

$$\chi = M^3 \, (C/Re)^{1/2}$$

where

$$C = \text{Chapman-Rubesin constant} = \rho_W \, \mu_W \, / \rho_e \, \mu_e$$

Subscript W indicates wall or surface value

Subscript e indicates value at outer edge of boundary layer.

Near the body leading edge or stagnation point ($\chi > 3$) there is a strong viscous interaction. Viscous interaction is weaker for smaller values of χ. Pressure and force coefficients (as well as skin friction and heat transfer coefficients) can be correlated with a modified viscous interaction parameter V

$$V = M \, (C/Re)^{1/2}$$

For hypersonic vehicles the maximum L/D decreases as the modified viscous interaction parameter increases. This is because the change in lift (with increase in V) is small compared with the increase in pressure and skin friction drag.

Another type of viscous interaction occurs when a shock wave intersects or impinges on a boundary layer; this is called shock-boundary layer interaction. The incident shock can cause local flow separation. The separated boundary layer may induce a second shock wave. When the boundary layer reattaches to the surface a third, or reattachment, shock wave may be formed (Anderson, 1989). In hypersonic flows, shock-boundary layer interaction can cause severe local aerodynamic heating problems.

Local heat fluxes can increase dramatically due to shock wave impingement. Hypersonic pressure and heat transfer effects in complex flow geometries can be extremely difficult to predict and experimental verification is normally required. NASA's Mach 6.7 experimental X-15 aircraft experienced severe heating caused by shock wave impingement and resulted in pylon failure and damage to the aircraft.

3.8   Heat Transfer

Structural materials, and materials used for thermal protection, have a maximum allowable temperature and can absorb a limited quantity of heat. The vehicle must also be designed to withstand the thermal stresses induced by temperature gradients as heat flux can vary along the vehicle surface. Structural strength decreases as temperature and heat flux increase. Excessively high heat fluxes can cause spalling (loss of surface material) or structural failure due to thermal stress.

Heat is transferred from the high-temperature shock layer to the vehicle by forced convection and by radiation. Some heat may be re-radiated to space. Heat is transferred by conduction through the structure to the vehicle interior. A circulating liquid or transpiration cooling system (injecting cold gas into the boundary layer) can dissipate some of this heat. An active cooling system may not be practical for an SRV due to added weight and complexity. If an ablative heat shield is employed, other modes of heat and mass transfer must be considered. (Ablation is discussed in sec. 6.)

3.8.1   Convective Heat Transfer

Many of the approximations for convective heat transfer in high-temperature gasses are based on the work of Fay and Riddell, 1958. The convective heat flux at the stagnation point $q_{CONV\,t}$ can be expressed as a function of freestream velocity V, freestream mass density ρ, and the body radius, or radius of curvature, r_N (Tauber, 1989)

$$q_{CONV\,t} \approx 1.83 \text{ x } 10^{-4} \, (\rho/r_N)^{1/2} \, V^3 \, [1 - (h_W \, / \, h_t)] \text{ W/m}^2$$

where

h_t = stagnation enthalpy = $c_p \, T + 1/2 \, V^2$ (for a perfect gas)

h_W = specific enthalpy at the vehicle surface

ρ is in kg/m^3, r_N is in m, V is in m/s, and h is in J/kg

Enthalpy is a function of wall temperature T_W and stagnation pressure p_t. Total enthalpy, or heat content, is the sum of internal energy and the product of volume multiplied by the pressure. Specific enthalpy, or enthalpy per unit mass, h is the sum of internal energy per unit mass and the ratio of the pressure divided by density (Liepmann and Roshko, 1957). Internal energy is the kinetic energy due to molecular motion (translational, rotational, and vibrational) and potential energy associated with the vibrational and electric energy of atoms within molecules. The change in internal energy is equal to the sum of the heat input to a system and the work done on or by the system.

The Tauber approximation given above is much simpler and usually more conservative than the more accurate Fay-Riddell solution.

For $h_W \ll h_t$, (cold-wall approximation) the maximum stagnation point convective heat flux $q_{CONV\,t\,MAX}$ for "steep" (flightpath angle $\gamma_E < -10$ deg), nonlifting entry is (Tauber, 1998)

$$q_{CONV\,t\,MAX} \approx 7.5 \times 10^{-7}\, V_E^3\, (BP \sin |\gamma_E| / r_N)^{1/2}\ W/m^2$$

where the ballistic parameter[26] BP is equal to $m/C_D\,S$ in kg/m^2 and γ_E is the flightpath angle at entry (see fig. 16).

Note that cold-wall heat flux is not a realistic prediction of actual flight conditions. Cold-wall heat fluxes should not be used for design purposes even though they may be useful for comparing designs.

For "steep," nonlifting trajectories, peak stagnation point convective heat flux occurs at altitude $y = (1/\beta) \ln [-\rho_0 / (\beta\, BP \sin \gamma_E)]$; ρ_0 and β are defined in section 2.

Note that for a given ballistic parameter and a given nose radius, the peak stagnation point heat flux decreases as the flightpath angle increases (negative flightpath angles point downward). Peak stagnation point heat flux is inversely proportional to the square root of nose radius and directly proportional to entry velocity cubed and the square root of ballistic parameter.

For $h_W \ll h_t$, the stagnation point convective total heat load $Q_{CONV\,t}$ for "steep," nonlifting entry is the integral of the heat flux

$$Q_{CONV\,t} \approx 0.028\, V_E^2\, BP^{1/2} / (r_N \sin |\gamma_E|)^{1/2}\ J/m^2$$

Stagnation point convective total heat load is inversely proportional to the square root of nose radius and directly proportional to entry velocity squared and the square root of ballistic parameter. For a given ballistic parameter and a given nose radius, stagnation point convective total heat load decreases as the flightpath angle becomes steeper. (Flight times decrease as the trajectory becomes steeper.)

A "normalized" cold-wall peak stagnation point convective heat flux[27] $q_{CONV\,t\,MAX}\,(r_N/BP)^{1/2}$ is given in table 7 and figure 17 as a function of entry velocity for various flightpath entry angles.

A "normalized" cold-wall stagnation point total heat load $Q_{CONV\,t}\,(r_N/BP)^{1/2}$ is given in table 8 and figure 18 as a function of entry velocity for various flightpath entry angles.

---

[26] In ballistics the quantity $W/C_D A$ is referred to as the ballistic coefficient and is usually given in English units, pounds per square foot. In this document the ballistic parameter BP is defined as $m/C_D\,S$ and is given in metric units, kg per square meter. (The reference areas S and A are equivalent.)

[27] A normalized quantity can be defined as a function or a variable that is multiplied or divided by one or more basic parameters. The normalized quantity is used here to scale the physical quantity (heat flux or total heat load) for various parameters such as vehicle mass, size, and shape (drag coefficient) and various initial conditions (entry velocity and flightpath angle at entry).

TABLE 7. "NORMALIZED" COLD-WALL PEAK STAGNATION POINT CONVECTIVE HEAT FLUX, $q_{CONV\,t\,MAX}\,(r_N/BP)^{1/2}$ IN kW/(kg-m)$^{1/2}$ FOR "STEEP" NONLIFTING ENTRY*

Entry Velocity, V_E km/s	Flightpath Angle at Entry, γ_E, deg					
	−10	−15	−25	−40	−65	−90
11	416	508	649	800	950	998
12	540	659	843	1039	1234	1246
13	687	838	1071	1321	1569	1648
14	858	1047	1338	1650	1959	2058
15	1055	1288	1646	2029	2410	2531

* $h_W \ll h_t$ (cold-wall approximation); $1/\beta = 7200$ m (see Tauber, 1998).

Example:
If $V_E = 15$ km/s and $\gamma_E = -25$ deg, then $q_{CONV\,t\,MAX}\,(r_N/BP)^{1/2} = 1646$ kW/(kg-m)$^{1/2}$.
If $r_N = 0.1$ m and BP $= 10$ kg/m^2, then $q_{CONV\,t\,MAX} = 1646\,(BP/r_N)^{1/2} = 1646\,(10/0.1)^{1/2} = 16{,}460$ kW/m$^2 = 1646$ W/cm^2.

TABLE 8. "NORMALIZED" COLD-WALL STAGNATION POINT CONVECTIVE TOTAL HEAT LOAD $Q_{CONV\,t}\,(r_N/BP)^{1/2}$ IN kJ/(kg-m)$^{1/2}$ FOR "STEEP" NONLIFTING ENTRY*

Entry Velocity, V_E km/s	Flightpath Angle at Entry, γ_E, deg					
	−10	−15	−25	−40	−65	−90
11	8130	6660	5210	4230	3560	3390
12	9680	7920	6200	5030	4240	4030
13	11,360	9300	7280	5900	4970	4730
14	13,170	10,790	8440	6840	5760	5490
15	15,120	12,380	9690	7860	6620	6300

* $h_W \ll h_t$ (cold-wall approximation); $1/\beta = 7200$ m (see Tauber, 1998).

Example:
If $V_E = 15$ km/s and $\gamma_E = -25$ deg, then $Q_{CONV\,t}\,(r_N/BP)^{1/2} = 9690$ kJ/(kg-m)$^{1/2}$.
If $r_N = 0.1$ m and BP $= 10$ kg/m^2, then $Q_{CONV\,t} = 9690\,(BP/r_N)^{1/2} = 9690\,(10/0.1)^{1/2} = 96{,}900$ kJ/m^2 or 9690 J/cm^2.

3.8.2   Radiative Heat Transfer

At Mars SRV Earth entry speeds the entry vehicle will be subjected to significant amounts of radiative heating from the high-temperature shock layer gasses; at these entry speeds the radiative heating can exceed the convective heating. High-speed radiative heat transfer is a complex, nonlinear process and numerical solutions are very difficult to obtain. A correlation can be used to predict the stagnation point radiative heat flux as a function of r_N, V and ρ (Tauber and Sutton, 1991):

$$q_{RAD\,t} = C\,r_N{}^A\,\rho^B\,f(V)$$

where

A is a function of ρ and V

B and C are constants

f(V) is a tabulated function for Earth's atmosphere

and r_N, V and ρ are as previously defined.

Note that the stagnation point radiative heat flux increases with increased nose radius. The opposite is true for stagnation point convective heat flux which decreases as nose radius is increased.

4. DYNAMICS AND FLIGHT MECHANICS

4.1 Shuttle Orbiter Re-entry

A sample return vehicle will enter the Earth's atmosphere at a much higher velocity than the Space Shuttle Orbiter.[28] However, the Orbiter is a familiar vehicle and it is useful to examine its re-entry from low Earth orbit (LEO) because it exhibits many characteristics similar to an SRV.

The Space Shuttle Orbiter orbits the Earth at an altitude of 320 km (200 mi) and a velocity of 7.7 km/s. This is in the free molecular flow regime where the molecular mean free path is large compared to a characteristic vehicle dimension. (The mean free path is defined as the average distance that gas molecules travel between successive collisions with one another; see section 2.1.3.) About 60 minutes before touchdown a de-orbit burn occurs. The re-entry event sequence is shown in table 9. The vehicle is turned tail forward and the Orbital Maneuvering System (OMS) engines are fired for 2 to 3 minutes. At about 160-km altitude, the vehicle leaves the free molecular flow regime. The vehicle continues to descend at hypervelocity speed and reaches the "entry interface altitude" (122 km) about 30 minutes after initiation of the de-orbit burn. About 5 minutes later the communications blackout begins. This is caused by ionized particles enveloping the vehicle. Transition from a laminar to a turbulent boundary layer occurs as Reynolds number increases above 1,000,000. (This value of transition Reynolds number is appropriate for smooth bodies with convex surfaces.) Maximum heating (nose and leading edge temperatures about 1500 °C) occurs about 20 minutes prior to touchdown. The flow around the vehicle is not in chemical or thermodynamic equilibrium and oxygen and nitrogen atoms, NO ions, and electrons are present. Continuum flow begins at an altitude of approximately 85 km and a Mach number around 25. Roll reversal (S-turn) maneuvers begin 16 minutes prior to touchdown. The communications blackout ends about 12 minutes prior to touchdown. Other high-temperature effects (e.g., nonequilibrium, molecular dissociation and vibrational excitation) become insignificant as the vehicle continues to decelerate. Braking begins 10 minutes prior to touchdown. At 6 minutes prior to touchdown the flow is no longer hypersonic ($M \approx 3$). At 3 minutes prior to touchdown the vehicle is traveling at sonic velocity (295 m/s at an altitude of 15 km); this is the brief transonic flow regime. At 2 minutes prior to touchdown the flow becomes subsonic ($M < 1$) and the vehicle is on a 22 deg glide slope. At 30 seconds prior to touchdown a flare maneuver begins and the glide slope is reduced to 1.5 deg. Landing speed is approximately 94 m/s. After the Orbiter has landed, cooling vans are dispatched to remove the heat accumulated in the TPS tiles during re-entry.

The flow regimes for a typical Space Shuttle Orbiter re-entry and landing are shown in figure 19. Note that the values and flow regime boundaries presented above and in figure 19 apply only to Orbiter-size vehicles and trajectories. Smaller vehicles would encounter free molecular flow at lower altitudes.

The important high velocity and high temperature effects can be summarized as follows (Hansen and Heims, 1958; Howe, 1990; and Anderson, 1989):

---

[28] The Space Shuttle Orbiter is also referred to as the Space Transportation System (STS) Orbiter.

The gas is calorically perfect at velocity < 1 km/s or temperature < 800 K

Vibrational excitation[29] occurs at velocity > 1 km/s (temperature > 800 K)

Oxygen dissociation occurs at velocity > 2.5 km/s (temperature > 2500 K)

Nitrogen dissociation and NO ionization occur at velocity > 5 km/s (temperature > 4000 K)

Ionization of atomic nitrogen and oxygen occurs at velocity > 10 km/s (temperature > 9000 K)

TABLE 9. SPACE SHUTTLE ORBITER RE-ENTRY EVENTS AND FLOW REGIMES

Re-entry Event and Flow Field Characteristics	Time before Touchdown, min	Velocity, km/s	Altitude, km
On-orbit, free molecular flow (hypervelocity)	—	7.7	320
De-orbit burn	60	7.7	282
Pitch up to ≈ 30 deg	52	7.6	—
Begin transition from free molecular to continuum flow	—	—	≈ 160
Entry interface, atmospheric effects begin	30	7.6	122
Blackout begins/NO ionization	25	7.4	95
Begin continuum flow regime, hypersonic flow (M ≈ 25)	—	—	≈ 85
Maximum heating (boundary layer transition/maximum deceleration)[a]	20	6.7	70
Dissociation of NO ends, roll reversal begins	16	≈ 5	—
Blackout ends	12	3.6	55
O_2 dissociation ends, braking begins	10	≈ 2.5	—
Vibration excitation ends, supersonic flow (M ≈ 3)	6	0.99	27
Transonic flow (M ≈ 1)	3	0.30	15
Subsonic flow (M < 1)	—	—	—
Hold 22 deg glide slope	2	0.19	4.0
Flare begins (1.5 deg glide slope)	0.5	0.16	0.6
Touchdown (or recovery)	0	0.09	0

[a] The values shown are for maximum heating. Boundary layer transition generally occurs earlier than maximum heating; peak deceleration generally occurs later than maximum heating.

---

[29] The vibrational energy of the molecules becomes excited and the specific heats and the isentropic exponent are no longer constant.

4.2 Trajectory Parameters and Vehicle Design Considerations

In 1953 Allen and Eggers[30] concluded in their study of ballistic missiles entering the Earth's atmosphere at high speeds (Allen and Eggers, 1958): "Blunt shapes appear superior to slender shapes from the standpoint of having lower maximum convective heat-transfer rates in the region of the nose." This became the basis for NASA's early spacecraft. The Mercury, Gemini and Apollo re-entry spacecraft had blunt, ablative heat shields.

Other conclusions about hypervelocity re-entry reached by Allen and Eggers include: a) high-drag (sphere-like) bodies should be selected to minimize convective heat transfer (large nose radius); b) vehicle weight, surface area, and skin friction drag should be minimized; c) vehicle surface should be stiff, strong and curved; d) aerodynamic force sensitivity to angle of attack should be minimized; and e) the body should be rotated in flight to equalize heat transfer to all surface elements, reduce temperature gradients, and achieve a more uniform surface temperature distribution.

Allen and Eggers recognized that high drag and low terminal speeds could result in excessive wind drift and high peak deceleration. They noted that these disadvantages could be alleviated with a variable geometry extendible skirt at the base of the vehicle to modulate drag.

SRVs and SRV trajectories must be designed to ensure survival of the vehicle payload after atmospheric entry. Special consideration must be given to maximum heat flux, total heat load (the integral of heat flux), and maximum deceleration. Peak structural loads generally occur at peak dynamic pressure or when the product of dynamic pressure and angle of attack reaches a maximum. These are some of the aerodynamic and aerothermodynamic factors that influence SRV design.

4.3 Equations of Motion

The equations of motion for a vehicle entering a planetary atmosphere can be written in inertial coordinates for a rotating planet. It is much simpler to assume planar motion over a nonrotating planet. For Earth the inertial velocity differs from the relative velocity by a westward-pointing vector of magnitude 462.8 cos λ m/s at sea level, where λ is the Earth latitude.[31]

If vehicle mass m is assumed constant (no thrust or mass loss due to ablation), then (Ashley, 1992)

$$m \, dV/dt = - D - m \, g \sin \gamma$$
$$m \, V \, d\gamma/dt = (1/r) \, m \, V^2 \cos \gamma + L - m \, g \cos \gamma$$
$$dr/dt = dy/dt = V \sin \gamma$$

where r is the distance from the center of the planet to the vehicle center of mass (see fig. 16).

---

[30] The original report was declassified in 1957 and published in 1958 as NACA Report 1381; see NACA-TR-1381 in the NACA report online database (National Advisory Committee for Aeronautics, 2006).

[31] Ashley (1992) claims errors due to neglecting Earth rotation are usually much less than 10 percent.

In terms of the ballistic parameter BP and the lift-drag ratio L/D:

$$dV/dt = -\rho \, V^2 / (2 \, BP) - g \sin \gamma$$

$$d\gamma/dt = (L/D) \, \rho \, V / (2 \, BP) - (g - V^2 / r) \cos \gamma / V$$

$$dy/dt = V \sin \gamma$$

where (see table 4 and section 2.4)

$$BP = m/C_D \, S$$

$$g = g_0 \, (1 + y / R_0)^{-2}$$

$$g_0 = 9.80665 \text{ m/s}^2$$

$$r = R_0 + y$$

$$R_0 = 6,356,766 \text{ m}$$

$$\rho \approx \rho_0 \, e^{-\beta y}$$

$$\rho_0 = 1.2250 \text{ kg/m}^3$$

$$\beta = 1/7200 \text{ m}^{-1}$$

Approximate analytical solutions of these equations can be readily obtained (Allen and Eggers, 1958 and Chapman, 1959 among others). These equations can also be solved by simple numerical integration for given initial conditions with BP and L/D assumed constant. Reasonably good results can be obtained by using a fourth-order Runge-Kutta method (Scarborough, 1966) with sufficiently small time steps (1 second or less). Runge-Kutta-Fehlberg methods vary the size of the time step depending on the behavior of the solution; step size is reduced if the solution varies rapidly (Fehlberg, 1969 and Forsythe et al., 1977).[32]

4.4 Maximum Deceleration for Steep Nonlifting Entry

For "steep" ($\gamma_E < -10$ deg), nonlifting entry (L/D = 0), with constant drag (BP = const), constant gravitational acceleration (g = g_0) and exponential atmosphere ($\rho = \rho_0 \, e^{-\beta y}$), the equations of motion can be solved for velocity V and acceleration dV/dt (Allen and Eggers, 1958)

$$V = V_E \, e^{\rho_0 e^{-\beta y} / (2\beta BP \sin \gamma_E)}$$

$$dV / dt = -\frac{\rho_0 V_E^2}{2BP} e^{-\beta y} e^{\rho_0 e^{-\beta y} / (\beta BP \sin \gamma_E)}$$

---

[32] Runge-Kutta-Fehlberg routines are available in commercial software application packages designed for scientific and engineering computations.

where

V_E = entry velocity, m/s

γ_E = entry angle (down is negative)

e = 2.7182818 (base of natural logarithms)

β = inverse scale height (exponential density approximation)

ρ_0 = 1.2250 kg/m^3

y = geometric altitude, m.

Maximum deceleration $- dV/dt|_{max}$ occurs at altitude $y = (1/\beta) \ln [-\rho_0/(\beta \, BP \sin \gamma_E)]$ and is a function of entry velocity V_E and flightpath angle at entry γ_E

$$-dV/dt\,|_{MAX} = \beta V_E^2 \sin \gamma_E / 2e$$

At peak deceleration

$$V = e^{-1/2} V_E \approx 0.61 \, V_E$$

Maximum deceleration for "steep" nonlifting entry is presented as a function of flightpath angle at entry and entry velocity in table 10 and figure 20. Extremely high decelerations (greater than 250 Earth g) are encountered for very steep entry angles ($\gamma_E < -50$ deg).

TABLE 10. DIMENSIONLESS MAXIMUM DECELERATION $(-dV/dt)/g$ FOR "STEEP" NONLIFTING ENTRY[a]

Entry Velocity, V_E, km/s	Flightpath Angle at Entry, γ_E, deg				
	−10	−30	−50	−70	−90
11	54	158	241	296	315
12	65	188	287	353	375
13	76	220	337	414	440
14	89	255	391	480	511
15	102	293	449	551	586

[a] g = 9.80665 m/s^2 and $1/\beta$ = 7200 m.

The altitude at which peak deceleration occurs is a function of the ballistic parameter and flightpath angle at entry; it is independent of entry velocity. Peak deceleration altitude decreases as ballistic parameter increases and as the flightpath angle at entry decreases.[33] Note that slender bodies (lower drag) usually have higher ballistic parameters than blunt bodies. The altitude for maximum

---

[33] This reference is to the actual value of the flightpath angle at entry, not to its absolute value. (A change of entry angle from −30 deg to −45 deg represents a decrease in entry angle.)

deceleration for "steep" nonlifting entry is presented as a function of flightpath angle and ballistic parameter in table 11 and figure 21.

TABLE 11. ALTITUDE IN KILOMETERS AT MAXIMUM DECELERATION FOR "STEEP" NONLIFTING ENTRY[a]

Ballistic Parameter, BP kg/m^2	Flightpath Angle at Entry, γ_E, deg				
	−10	−30	−50	−70	−90
5	66.4	58.8	55.7	54.3	53.8
10	61.4	53.8	50.7	49.3	48.8
25	54.8	47.2	44.1	42.7	42.2
50	49.8	42.2	39.2	37.7	37.2
100	44.9	37.2	34.2	32.7	32.3
200	39.9	32.3	29.3	27.7	27.3

[a] $\rho_0 = 1.2250$ kg/m^3 and $1/\beta = 7200$ m

4.5 Lifting and Shallow Entry

Chapman found a simple and more general solution to the equations of motion by introducing a coordinate transformation (Chapman, 1959). Numerical solutions have been tabulated for the following conditions (Chapman and Kapphahn, 1961):

$$8 \text{ km/s} < V_E < 24 \text{ km/s}$$

$$-70 \text{ deg} < \gamma_E < 0$$

$$-4 < L/D < 4$$

Chapman and Kapphahn's solutions can be used for nonlifting entry (L/D = 0) and for lifting entry (negative as well as positive lift-drag ratios).

Chapman was the first U.S. scientist to study the atmospheric entry corridor problem (see Chapman, 1960). If a ballistic entry vehicle enters the atmosphere at too shallow an angle, it will skip out of the atmosphere. If the flightpath entry angle is too steep, the vehicle may burn up or suffer damage due to excessive structural loads. Guidance accuracy requirements are also discussed in Chapman's work. Temperature constraints are discussed in section 6 (see Kolodziej, 1997).

4.6 Detailed Entry Trajectory Calculations

Detailed trajectory simulations can be performed using three- or six-degree-of-freedom trajectory codes. The Program to Optimize Simulated Trajectories (POST) code has been used for this purpose for many years. POST does three degree-of-freedom (DOF) simulation. There is also a six-degree-of-freedom version called 6D POST.[34] POST has been used successfully to solve a wide variety of atmospheric entry problems. The program is capable of simulating and optimizing trajectories for space vehicles.

Trajectory simulation codes include various atmospheric and gravity models. Aerodynamic coefficients (Newtonian, free molecular, flight- or ground-test data or CFD results) and aerodynamic heating models (both laminar and turbulent) are usually included as well. Some codes include propulsion models and reaction control system models. Navigation, guidance and flight controls can be simulated as well.

Trajectory simulations are used at various stages of the design process. Generally, this is accomplished by selecting: a) appropriate atmosphere and gravity models; b) vehicle size, shape, mass, center of mass, TPS material, propulsion model; and c) initial conditions (entry coordinates, velocity, flightpath angle, and vehicle attitude).

The following trajectory output results are obtained: a) altitude, velocity and acceleration profiles; b) vehicle attitude and stability; c) aerodynamic loads, M, Re, and dynamic pressure; d) heat transfer rates and heat loads; e) vehicle thermal response (heatshield and structure temperatures); and f) impact location.[35]

Results are evaluated to determine if mission requirements and constraints are satisfied. If not, vehicle size, shape, and material characteristics, as well as initial conditions, are modified until a satisfactory design is obtained.

A trajectory simulation code called TRAJ (Allen et al., 2005) can be used for 3- and 6-DOF trajectory simulation and thermal protection system design.[36] The TRAJ code has been validated with flight and ground test data for both trajectory variables and heatshield thermal response. Examples of TRAJ results for Stardust entry are presented in sections 6 and 8 (see figs. 30 through 34).

---

[34] At one time many software packages developed by government agencies and by government contractors were available to industry and to the public at cost from COSMIC or from NASA's Software Technology Transfer Center. Many of these items can be obtained from Open Channel Software (Open Channel Foundation, 2006). The POST code is described in NASA contractor reports (Brauer et al., 1975).

[35] Monte Carlo statistical simulations can be used to determine downrange and crossrange impact dispersions. This requires knowledge of variations in SRV properties, dimensional tolerances, accuracy of approximations, atmospheric temperature profiles, winds, and several other variables. It is necessary to run a very large number of trajectories to get accurate results.

[36] Copies of the TRAJ code are available from NASA with approval of an export control official. Point of Contact: Gary A. Allen, Jr., gallen@mail.arc.nasa.gov, 1-650-604-4228, MS 230-3, NASA Ames Research Center, Moffett Field, CA 94035-1000.

4.7 Stability

Vehicle stability during atmospheric entry is not a simple matter. The vehicle flow field is complicated and involves several flow regimes (from free molecular to subsonic). It is often necessary to perform full six-degree-of-freedom (6-DOF) trajectory simulations even for simple vehicle geometries. Six-DOF trajectory simulations require knowledge of stability derivatives, mass properties (weight and CG locations) and moments of inertia. Stability derivatives are usually determined experimentally. An additional equation of motion is required to account for rotation:

$$I\ddot{\theta} = \Sigma M = q\,S\,L\,\Sigma\,C_m$$

where

q = dynamic pressure

S = reference area

L = reference length (for pitching moment coefficient)

I = pitch moment of inertia about the center of gravity

θ = vehicle pitch angle measured from local horizontal

M = pitching moment

C_m = pitching moment coefficient = $M / (q\,S\,L)$.

and Σ indicates the sum of the moments about the center of gravity.

In the hypersonic speed range the vehicle is usually statically stable (pitch oscillation amplitude decreasing as altitude decreases) if the static margin is adequate (center of gravity forward of center of pressure). However, it may be necessary to spin the vehicle to provide additional dynamic stability. The vehicle may also experience coning and tumbling motions.

4.7.1 Static Stability

For steady, horizontal flight, the pitching moment M about the vehicle center of mass (see fig. 12) is

$$M = -N(x_{CP} - x_{CG})$$

$$C_m\,qSL = -C_N\,qS(x_{CP} - x_{CG})$$

$$C_{m\alpha} = -C_{N\alpha}(x_{CP} - x_{CG})/L$$

where

$$C_N = N/qS$$

$$C_{m_\alpha} \equiv \frac{\partial C_m}{\partial \alpha} = \text{pitching moment curve slope}$$

$$C_{N_\alpha} \equiv \frac{\partial C_N}{\partial \alpha} = \text{normal force curve slope.}$$

The vehicle is stable if C_{m_α} is less than zero <u>and</u> if the center-of-gravity location x_{CG} is forward of the center-of-pressure location x_{CP}:

$$C_{m_\alpha} < 0 \text{ and } x_{CP} - x_{CG} > 0$$

The quantity $(x_{CP} - x_{CG})$ is often referred to as the static margin. The larger the value of the static margin the more stable the vehicle in the sense of static stability (Etkin, 1972).

A sharp, conical body in steady horizontal, flight will be statically stable in the pitch plane if C_{m_α} is negative and unstable if C_{m_α} is positive. The Newtonian approximation for the pitching moment curve slope for sharp, conical bodies is shown in figure 22. Figure 22 also shows the effect of CG location x_{CG}. A sharp conical body becomes less stable as the CG location moves aft (closer to the downstream-facing base of the cone). For a sharp cone at zero angle of attack, the CP location x_{CP} is given by Newtonian theory

$$x_{CP} = 2/3\, L\, (1 + \tan^2 \theta_C)$$

where L = cone length and θ_C = cone semivertex angle.

4.7.2   Dynamic Stability

In atmospheric entry the motion of an entry vehicle may not be stable. As altitude decreases dynamic pressure first increases to a maximum and then decreases to impact (or until terminal velocity is reached). The increase in dynamic pressure results in pitch oscillations if the vehicle axis is not aligned with the velocity vector at entry.

Tobak, Allen, and Sommer studied hypersonic dynamic stability for atmospheric entry at Ames in the 1950s. They established dynamic stability criteria and analytically determined the amplitude and frequency of pitch oscillations (Tobak and Allen, 1958 and Sommer and Tobak, 1959). They also derived an expression for pitch angle as a function of altitude in terms of Bessel functions (Allen, 1957).

An application of Tobak's method is to determine the dynamic stability parameter as a function of altitude and estimate the convergence-divergence boundary for unstable pitch oscillations (i.e., the altitude at which the sign of the dynamic stability parameter changes from positive to negative). Table 12 shows the results of this calculation for three flightpath entry angles. In each case pitch oscillations due to finite angle of attack at entry converge at altitudes above the convergence-divergence altitude shown in table 12 and diverge at lower altitudes. Note that these results are valid only for "steep" entry angles and hypersonic speeds.

TABLE 12. CONVERGENCE-DIVERGENCE ALTITUDE FOR PITCH OSCILLATION STABILITY FOR "STEEP" NONLIFTING ENTRY (BP = 50 kg/m^2, V_E = 15 km/s AND y_E = 125 km)[a]

Flightpath angle at entry, γ_E, deg	Convergence-divergence altitude, km
−20	44.9
−40	40.4
−60	38.3

[a] For L/D = 0 and M >> 1. Note that these results cannot be extrapolated to altitudes below 34 km.

5. DECELERATION AND RECOVERY

In general, deceleration and recovery planning begins with trajectory computations (see sec. 4). The simplest case is nonlifting entry where the critical variable, the ballistic parameter[37] BP is constant. More complicated cases arise when constant or variable lift is introduced to alter the trajectory. Lift can be used to change orbit plane, reduce heat flux or total heat load, or improve targeting accuracy. It is also possible to control the aerodynamic drag force acting on the vehicle; this is called "drag modulation."

5.1 Drag Modulation

Drag modulation can be used to change the speed of a nonlifting vehicle. Drag increases as the vehicle cross-sectional area (in a plane normal to the velocity vector) increases. If lift is assumed equal to zero, increasing or decreasing the vehicle cross-sectional area can modulate drag. Drag modulation can be used to constrain the trajectory variables and control the impact location. However, drag modulation is less effective than varying the lift of the vehicle (lift modulation).

A drag modulation concept was proposed for an Aero-Assist Orbital Transfer Vehicle (AOTV) in the 1980s. This concept was a combination of a balloon and a parachute and termed a "ballute." Unlike wake-deployed ballutes the AOTV ballute inflated around the vehicle. Drag modulation would occur by increasing or decreasing the pressure inside the balloon, which would in turn change the cross-sectional area of the vehicle. The concept attracted much interest due to its apparent simplicity. However, it was difficult to implement and the concept was eventually dropped. The major difficulty was providing thermal protection for the inflatable structure. The Soviet Union has used inflatable braking devices for planetary entry in the past and Russia continues to study their use.

5.2 Parachute Recovery

For entry vehicles with low values of L/D at subsonic speeds the use of a parachute for final descent and landing is an attractive option. This approach has been used for blunt entry capsules (both Earth and planetary; Apollo and Mars Pathfinder, for example) and also proposed for lifting vehicles with poor subsonic flight qualities. Mach number and dynamic pressure, along with Reynolds number, are the critical variables. Lower Mach numbers are preferred as it is difficult to deploy a parachute at high supersonic Mach number and high dynamic pressure. The problems include: deployment shock to the material; uncertainty of proper parachute inflation; and difficulty of predicting overall performance (particularly in the transonic-supersonic range where shock waves and uncertain pressure distributions are encountered).

---

[37] The ballistic parameter BP is defined here as $m/C_D S$ and its units are kg/m^2. A similar quantity in English units is called the ballistic coefficient and is defined as $W/C_D A$ and given in units of lbm/ft^2, where W is the body weight and the reference area A is equivalent to the area S in the definition of ballistic parameter.

For most applications, parachute deployment usually occurs around Mach 1.8 because of dynamic instability at lower Mach numbers. Since the flow field is greatly influenced by body shape ground or flight testing is usually required.

5.3   Final Descent and Landing

A heatshield having significant thermal mass can transfer thermal energy to the interior of a SRV. Even though the heatshield surface temperature decreases rapidly as velocity decreases at lower altitudes heat conduction from the heatshield to the vehicle interior can continue after landing. For Space Shuttle Orbiter flights structural temperatures beneath the TPS tiles would continue to increase after the vehicle had landed if external cooling was not applied.

During the final descent of the Mars Pathfinder the forward heatshield was separated from the lander vehicle. This method could also be employed on a Mars SRV entering the Earth's atmosphere. The Mars Pathfinder deployed a drogue chute and air bags that inflated just prior to impact.

6. HEAT TRANSFER AND THERMAL PROTECTION MATERIALS

6.1 Heat Transfer

Heat transfer prediction is critical to the design of a thermal protection system (TPS). Heat flux, total heat load and surface temperature histories affect the selection of: an appropriate heat-shield material; the heat-shield thickness; and the means of attaching the heatshield to the vehicle structure. In the 1950s, it was thought that there was a "thermal barrier," much like the belief that there was a "sonic barrier," that was impossible to go beyond. It was believed that the very high kinetic energy associated with an entry vehicle would greatly exceed the heat of vaporization of most materials. However, only a small fraction of the vehicle kinetic energy goes into actually heating the vehicle. Most of the energy is used to heat the air ahead of the vehicle as it is compressed by the bow shock wave. The thermal barrier may be readily penetrated with appropriate TPS design. (Allen's discovery that bluntness reduces the effect of entry heating is discussed in sec. 4.2.)

6.2 Radiative Heat Transfer

The dominant form of entry vehicle heat transfer is convective heat transfer at the vehicle surface. In some situations radiative heating from the shock layer gasses can be very significant. At vehicle velocities above 13 km/s in the earth's atmosphere, the shock layer becomes extremely hot (see fig. 2). At these velocities radiation from the shock layer to the vehicle surface can become a significant fraction of the total incident energy. Computational methods for radiative heat transfer effects are extremely difficult and become more difficult with the introduction of ablation products into the boundary layer. An added complexity is that thermodynamic and transport properties are not well understood at temperatures above 9000 K.

Flight testing may be necessary to determine radiative heat flux at conditions that would be experienced by an SRV. In some situations simple approximations for radiant heat flux q_{RAD} can be used. At the stagnation point on the body:

$$q_{RAD\,t} = C\, r_N^{\,A}\, \rho^B\, f(V)$$

where: A is a function of ρ and V; B and C are constants; and $f(V)$ is a function of velocity for a given atmosphere (Tauber and Sutton, 1991).

Radiative heating was a significant portion of the total heat flux for the Galileo Jupiter entry probe. For the Apollo Command Module approximately 34 percent of the total heat flux was due to radiative heating (see table 13). For a typical nonlifting entry body as much as 83 percent of the total heat load can be due to radiative heating (see table 13).

TABLE 13. PEAK STAGNATION POINT HEAT FLUX AND TOTAL HEAT LOAD FOR EARTH ENTRY SPACECRAFT

Entry vehicle	BP, kg/m^2	V_E, km/s	γ_E, deg	$q_{MAX\,t}$, W/cm^2	Q, J/cm^2
Apollo CM, L/D ≈ 0.3	≈ 500	≈ 11	—	≈ 510[b]	—
Stardust[a]	68.2	12.9	−8.2	856	23,730[c]
Entry Body,[a] L/D = 0	50	15	−10	1750	19,340[d]
Entry Body,[a] L/D = 0	50	15	−50	11,170	20,300[e]

[a] Results of NASA TRAJ simulation, 1999.
[b] Approximately 34 percent of the total heat flux is radiative and 66 percent is convective.
[c] The total heat load is 9 percent radiative and 91 percent convective.
[d] The total heat load is 61 percent radiative and 39 percent convective.
[e] The total heat load is 83 percent radiative and 17 percent convective.

The TRAJ simulation results for a body shaped like the Stardust entry capsule is shown in figures 23 and 24. At entry velocities less than 12 km/s the stagnation point heat flux and the stagnation point heat load are dominated by convective heating. As the entry velocity increases the radiant heat flux increases rapidly. At V_E = 15 km/s, the peak stagnation point radiative heat flux is more than two times greater than the peak stagnation point convective heat flux. At entry velocities greater than 15 km/s, the heat load due to radiation is greater than the heat load due to convection.

6.3   Heat Transfer at Vehicle Surface

At the surface of an entry vehicle heat is transferred from the hot gases in the shock layer to the vehicle surface by forced convection q_{CONV} and through the vehicle structure to the vehicle interior by conduction q_{COND}. The convective heat flux can be estimated using the methods given in section 3.8. Methods for estimating conductive heat flux are given in textbooks on heat transfer (Jacob, 1949). The Stefan-Boltzmann radiation law gives the re-radiative heat flux[38] at the vehicle surface:

$$q_{RE\text{-}RAD} = S\,\varepsilon\,\sigma\,(T_W^4 - T_{SPACE}^4)$$

where

> S = surface area
>
> ε = emissivity of material surface
>
> σ = Stefan-Boltzmann constant = 5.67040 x 10^{-8} W/(m^2-K^4)
>
> T_W = wall temperature
>
> T_{SPACE} = temperature of space or radiation sink.

---

[38] This is the heat transferred from the vehicle surface to the space environment by radiation.

Note that the fourth power of T_W shows that re-radiation is far more significant at high temperature. For practical considerations, it is desirable to minimize heat conduction to the vehicle interior (which is why low-density ceramics are often used) and to maximize the heat re-radiation (which is why surface coatings with high emissivity are desirable). Values of emissivity for a variety of TPS materials are given in NASA's TPSX online database (Squire, 2006).[39]

6.4 Heatshield Design Considerations

In designing a heatshield for an atmospheric entry vehicle, there are a variety of factors to consider. Two dominant factors are system weight and backwall design temperature; the latter determines how much thermal energy will flow to the vehicle interior. For large reusable vehicles, the economics of the system, as well as qualification costs, are of considerable concern. As an example, the Chinese have successfully used oak wood for an entry capsule TPS. The oak performed well and is economical but results in a fairly heavy vehicle. Lightweight ceramic materials have much lower thermal conductivity and much lower density than oak, but are much higher in cost. The payoff for lightweight TPS materials is the added mass that can be used for scientific instruments or other payload. SRV missions have large propellant mass requirements. If TPS mass can be reduced, less propellant will be required (or payload mass can be increased).

A method for heat shield design can be described as follows. First, the mission requirements are defined and trajectory calculations are made for specific entry conditions. Then first order calculations are made for critical heating areas on the vehicle at points along the flightpath. Thermal analyses are performed to determine required TPS thicknesses at critical locations. For TPS materials attached to the vehicle structure with high-temperature adhesives, the backwall temperature cannot exceed the maximum allowable temperature of the adhesive. Different materials can be selected for different locations on the vehicle. For example, rigid ceramic tiles can be used on windward (or leading edge) surfaces and flexible blanket materials such as advanced flexible reusable surface insulation (AFRSI) or tailorable advanced blanket insulation (TABI) can be used on leeward surfaces. At a later stage, a more detailed flowfield analysis of the vehicle can be performed to achieve optimal sizing. Analytical and numerical methods have been developed by NASA to greatly expedite this design process. Various trajectory simulation codes can be used for TPS sizing.

6.5 Methods of Thermal Protection for Sample Return Vehicles

Thermal protection system design is critical for sample return vehicle mission success. SRVs entering the Earth's atmosphere at velocities greater than 10 km/s will encounter severe heat fluxes and heat loads as shown in table 13. Methods for estimating convective and radiative heat fluxes for atmospheric entry vehicles are discussed in section 3.

An effective thermal protection system is required to assure survival of sample return capsules and to ensure that capsule interior temperature limits are not exceeded. An insulating material may be used to protect the load-bearing structure or an active cooling system may be employed. In severe heating

---

[39] Registration is required; some restrictions apply.

environments the insulating material may be allowed to ablate. Some common thermal protection schemes are described below.

6.5.1 Heat Sink

A heat-sink structure is a simple, passive thermal protection method.[40] It can be a homogeneous or composite material or a honeycomb structure. Composites are made up of two or more constituents bonded together to optimize thermal and/or mechanical properties. Composite materials are generally anisotropic (properties have different values when measured in different directions). In many applications heat-sink materials provide inadequate strength, excessive weight or uncertain survivability when subjected to entry heating. (Beryllium-copper was used as a heat-sink material for early re-entry vehicles, but fabrication was very difficult.)

6.5.2 Radiative Surfaces

Radiative surfaces dissipate a portion of the incident radiant energy by re-radiation. If the structure can tolerate radiation equilibrium temperatures, which vary as $q^{1/4}$, and the material has a low thermal conductivity, then radiative cooling can be very effective. The ratio of emitted radiant flux density to incident radiant flux density is called emissivity ε.

6.5.3 Film and Transpiration Cooling

Film or transpiration cooling can be used to protect vehicles from high entry-heating environments. In film cooling, a fluid is injected into the boundary layer near the stagnation point to protect the surface. In transpiration cooling, the fluid is injected through pores in the material surface. Film and transpiration cooling are less effective when the heat input is predominantly radiative rather than convective. Analyses and experiments show that the heat flux can be reduced significantly by increasing the mass addition rate (Tauber, 1989). The effectiveness of film or transpiration cooling can be further increased by selecting a coolant gas with a molecular weight lower than that of the boundary layer gases. Film and transpiration cooling are more effective in laminar flow than in turbulent flow. A disadvantage of film and transpiration cooling is that a complex supply and regulating system may be required. However, film or transpiration cooling may be required to protect small areas with sensors or windows.

6.5.4 Ablation

Ablation cooling is a very common method of thermal protection for planetary entry vehicles (e.g., Apollo Command Module, Mars Pathfinder and Galileo). Ablation materials are consumed as they are heated to sublimation temperature. Sublimation phase change (solid state to liquid or vapor) and gaseous transpiration promote surface cooling (material surface temperature affects the amount of heat radiated from the surface). Ablation is a complex process and can involve a variety of chemical reactions including: combustion; sublimation; vaporization (liquid to gas); erosion; oxidation; charring; pyrolysis; melting; molecular dissociation; and recombination. Figure 25 is a schematic diagram of the charring ablation process.

---

[40] A heat sink is a body that can absorb or reject a large quantity of heat without an appreciable change in temperature. A heat sink is also referred to as a heat reservoir.

NASA defines ablation as (Allen, 1965): *The removal of surface material from a body by vaporization, melting, chipping, or other erosive process; specifically the intentional removal of material from a nose cone or spacecraft during high speed movement through a planetary atmosphere to provide thermal protection to the underlying structure.*

The following description appears in the current online version of the NASA's Dictionary of Technical Terms (Glover, 2006): *Ablating materials are used on the surfaces of some reentry vehicles to absorb heat by removal of mass, thus blocking the transfer of heat to the rest of the vehicle and maintaining temperatures within design limits. Ablating materials absorb heat by increasing in temperature and changing in chemical or physical state. The heat is carried away from the surface by a loss of mass (liquid or vapor). The departing mass also blocks part of the convective heat transfer to the remaining material in the same manner as transpiration cooling.*

As shown in figure 25, vaporization and sublimation occur at the outer surface of the ablation material. Oxidation and combustion also occur at the outer surface and result in surface recession as the ablation material is consumed. The ablating surface may serve as a catalyst to promote recombination of oxygen and nitrogen atoms in the boundary layer. Pyrolysis, a chemical change caused by heat addition, begins as thermal energy penetrates into the material (at around 670 K for phenolic resins). About one-half of the material volatizes and the other half is left as a carbonaceous residue or char. Decomposition gases from the pyrolysis zone percolate through the char layer and absorb thermal energy.

Wood is an example of a charring ablation material. Synthetic resins cured with reinforcing fibers are used as charring ablators for entry vehicle heat shields. The energy absorbed per unit mass ablated is a measure of ablation effectiveness. For silicon nitride, the energy absorbed per unit mass ablated is 6 MJ/kg (Tauber, 1989). Properties of typical TPS materials are given in table 14.

The carbon-phenolic materials used on the Galileo probe may be too heavy (1200 to 1920 kg/m^3) for SRVs. The heat shield for the Apollo command module was made from AVCOAT-5026 (529 kg/m^3). Superlight ablators (SLAs) were used on the Mars Viking spacecraft in 1976 and the Mars Pathfinder spacecraft in 1997. SLA-561V has a density of 265 kg/m^3. A disadvantage of SLA materials is their high manufacturing cost and fragility. Phenolic impregnated carbon ablator (PICA) is a lightweight ceramic ablator (LCA) material. PICA can be made in densities from 225 to 880 kg/m^{3}and was selected for the Stardust sample return mission.[41]

Ablation material sizing can be performed using the Fully Implicit Ablation and Thermal Response (FIAT) code developed by Y.-K. Chen (Chen and Milos, 1997). The FIAT code is incorporated in NASA's TRAJ trajectory simulation code (Allen et al., 2005). The code is easy to use, but only four ablation materials are available. The user first specifies the vehicle geometry and initial conditions (y_E, V_E, and γ_E) and selects the heatshield material and thickness, then runs the trajectory simulation. The code calculates heatshield mass, surface and in-depth temperature-time histories, surface recession and consumable mass flux for the calculated trajectory. Another option is to allow the code to calculate an optimal heatshield thickness for a specified backface temperature. The second option does not always provide solutions. Typical results for a Stardust sample return mission are given in

---

[41] PICA is composed of a fibrous carbon substrate (manufactured by Fiber Materials, Inc.) and a phenolic transprant. PICA shows high effective heat of ablation at heat fluxes between 300–1500 W/cm^2.

table 15. Typical results for a nonlifting sample return entry vehicle are given in table 16. (Note that these results were obtained using a 1999 version of the TRAJ code.)

TABLE 14. PROPERTIES OF SOME TPS MATERIALS*

Material	Density, kg/m^3	Thermal Conductivity,[a] W/m-K	Emissivity,[a] ε	Melt Temp., K
AVCOAT 5026 (virgin)	529	0.242	0.67	1920
AVCOAT 5026 (char)	264	—	0.49	—
Carbon-Phenolic (Acusil I)	480	0.112	0.95	2000
Reinforced Carbon-Carbon (RCC)	1580	5.05 (normal) 7.88 (parallel)	0.54 to 0.9	2030[b]
Teflon	2190	0.251	—	—
Shuttle Tile (LI-900)	144	0.047	0.88	1755[b]
Beryllium	1840	170	—	810[b]
Oak	610	0.146	—	—

* Squire, 2006 (TPSX database).
[a] Thermal conductivity and emissivity are functions of temperature.
[b] Single-use temperature limit.

TABLE 15. TRAJ PLUS FIAT ABLATION SIMULATIONS FOR STARDUST SRV

Material	Q, kJ/cm^2	Q_{RAD}, kJ/cm^2	Heatshield Thickness, cm	Heatshield Mass, kg	Surface Recession, cm	Mass Loss, kg
PICA[a]	27	2	5.08	—	1.07	1.06
PICA-15[b]	23.7	2.1	5.08	7.55	0.76	0.26[c]
Carbon-Phenolic[b]	23.7	2.1	3.81	33.97	0.12	—

[a] FIAT simulation (Chen and Milos, 1998). Bondline (ablation material backface) temperature $\leq$ 523 K.
[b] TRAJ plus FIAT simulation, 1999. Initial conditions: y_E = 132.9 km, V_E = 12.86 km/s, γ_E = –8.2 deg for BP = 68.2 kg/m^2.
[c] Mass loss estimated from integral of char plus gas mass flux.

TABLE 16. TRAJ PLUS FIAT ABLATION SIMULATION FOR A NONLIFTING SRV[a]

Material	Q, kJ/cm^2	Q$_{RAD}$, kJ/cm^2	Heatshield Thickness, cm	Peak Surface Temp., K	Bondline Temp., K at t = 94 s	Surface Recession, cm
PICA-15	28.0	4.4	3.43	3536	440	1.36
Carbon-Phenolic	28.0	4.4	2.54	3262	439	0.18

[a] Initial conditions: y = 125 km, V_E = 15 km/s, γ_E = –10 deg for BP = 50 kg/m^2 (Blunt cone: θ_C = 25 deg, r_N = 0.171 m, r_B = 0.43 m). Initial bondline (ablation material backface) temperature = 239 K.

6.6 Ablation Constraint Determination

Kolodziej has developed a spreadsheet method to calculate atmospheric-entry constraints (Kolodziej, 1997). If vehicle characteristics are specified, a bounding altitude-velocity curve can be obtained that indicates a limit for a system variable (such as stagnation point temperature). For example, if it is necessary that the stagnation point temperature not exceed the material ablation temperature, then a bounding curve can be determined that separates the region where ablation occurs from the no-ablation region as shown in figure 26. For this example ablation occurs when the trajectory curve falls below or to the right of the "ablation-constraint" curve. This method includes hot wall, wall catalysis, imperfect gas properties and rarefied flow effects. An example is presented below.

For materials that are good insulators and radiate efficiently:

$$q_{CONV} - q_{COND} = \sigma \, \varepsilon \, T_W^4$$

If $q_{CONV} \gg q_{COND}$, $h_W \ll h_t$ and $T_W = T_{MAX}$ (the maximum nonablating use temperature of the material), then for an axially-symmetric body with nose radius r_N

$$q_{CONV} = \sigma \, \varepsilon \, T_{MAX}^4$$

and

$$q_{CONV} = const \; \rho^{1/2} \, V^3 \, / \, r_N^{1/2}$$

or solving for $V = V_{CONSTRAINT}$

$$V_{CONSTRAINT} = (\sigma \, \varepsilon / 0.000183)^{1/3} \, (r_N/\rho)^{1/6} \, T_{MAX}^{4/3}, \; m/s$$

If freestream density $\rho = \rho_0 \, e^{-\beta y}$ (where ρ_0 = 1.225 kg/m^3 and 1/β = 7200 m), then $V_{CONSTRAINT}$ is a function of altitude alone for given values of ε, T_{MAX}, and r_N. Ablation-constraint curves are shown in figures 27 through 29. The curves show the effects of emissivity, ablation temperature and nose radius on the ablation constraint. These curves correspond to values that might be expected for a noncatalytic cold wall. For a noncatalytic hot wall, the method shows the constraint is less

53

conservative (the curve is shifted down and to the right). For fully catalytic walls the constraint curve falls between the hot-wall noncatalytic and the cold-wall noncatalytic curves.

6.7 Hypersonic Simulation and Optimization Codes

The Hypersonic Vehicle Optimization Code (HAVOC) can be used to obtain estimates of TPS requirements (thickness, weight and cost) for a specified TPS material, vehicle geometry and trajectory. The HAVOC engineering design code has been used for sizing of nonablative TPS materials for hypersonic aircraft (Kolodziej et al., 1998).

An aerodynamic heating and TPS sizing code (MINIVER) can be used both to predict the aerothermal environment and to perform simple TPS sizing for aerospace vehicles that operate in the hypersonic flight regime (Wurster et al., 1999). The code uses approximate heating methods for predicting heat flux and perfect-gas or equilibrium-air chemistry for aerodynamic flow field approximations.

MINIVER has been used as design tool in government and industry for a wide variety of vehicle configurations. The code has the ability to quickly provide the thermal environments required for TPS sizing. MINIVER was originally developed at McDonnell Douglas in the 1970s under government contract. Many government, military, educational, and aerospace industry installations currently use it.[42]

6.7.1 Aerothermodynamics

Analytical methods are used to determine aerodynamic coefficients for various geometric configurations. Experimental data can be used to improve the results. In the supersonic/hypersonic speed range, lift and pressure drag are computed using tangent wedge/tangent cone methods. Hypersonic pressure drag is found using Newtonian theory. HAVOC includes base pressure drag, skin friction drag, and approximations for heat transfer coefficients. At present MINIVER does not include provisions for radiative heat transfer.

6.7.2 Structural Analysis

The HAVOC code has structural analysis and weight estimation capabilities. After body stress distributions are calculated, the minimum amount of structural material required to prevent failure is determined. Tensile yield, compressive yield, local buckling, and gross buckling failure modes are considered. TPS thickness determination is based on quasi-steady state, one-dimensional heat conduction analysis and the requirement to keep the interior of the vehicle below a specified temperature.

---

[42] NASA Point of Contact: Kathryn Wurster, NASA Langley Research Center, Mail Stop 451, Hampton, VA 23681-2199, 1-757-864-4487, K.E.Wurster@nasa.gov.

6.7.3 Trajectory Simulation

A three-degree-of-freedom trajectory simulation is available in the HAVOC code or a Mach number versus altitude profile can be specified. The equations of motion are for a point-mass body moving relative to a rotating, spherical earth. The trajectory can be constrained by maximum dynamic pressure or by maximum structural temperature.

7. GROUND AND FLIGHT TEST SIMULATION

A variety of analytical and experimental methods are available for use in SRV design. These methods may include a combination of ground tests to simulate portions of the flight regime or aspects of flow phenomena (e.g., aerodynamic loads, TPS thermal response or chemical reactions). As the flight concept matures simplified flight experiments can be performed to add important design information that is not available from ground facilities or computation (e.g., boundary layer transition at hypersonic velocities). The final validation is of course the actual flight of the SRV. Some of the critical simulation parameters (Mach number, Reynolds number, Knudsen number, etc.) were discussed in section 1. CFD methods are used extensively to determine aerodynamic and aerothermodynamic flow characteristics. However, ground- and flight-testing are still necessary to validate and "calibrate" the CFD methods and their results.[43]

An inventory of major government aerospace test facilities is available on the Internet (NASA Headquarters, 2006).

7.1 Ground Test

Aerodynamic ground testing facilities and techniques are described by Lukasiewicz (1973) and by Pope and Goin (1965). Brief descriptions of the principal types of ground-test facilities are given below.

7.1.1 Conventional Wind Tunnel

Conventional wind tunnels are used to determine aerodynamic forces and moments as well as vehicle surface pressure distributions. Compressibility and viscous effects are simulated by matching test Mach number and Reynolds number with flight values. For very high altitudes, Knudsen number is matched in low-density wind tunnels. Strain gauge balances are used to measure forces and moments. Pressure transducers are connected to pressure ports to measure surface static pressures. Nonintrusive velocity measurements are made with a laser Doppler velocimeter. Hot-wire anemometers are used to measure turbulent velocity fluctuations. There are a variety of flow visualization techniques to determine shock wave shape and location, surface streamline patterns and heat transfer characteristics[44] (Liepmann and Roshko, 1957).

At high Mach numbers (above around Mach 7) conventional wind tunnels are limited by test section size and run time. It is usually necessary to use small models and rapid-response data acquisition systems. There are also limitations on maximum available stagnation temperatures and pressures.

---

[43] In cases where there is little or no previous CFD experience, it is often necessary to vary the initial conditions and/or the turbulence model to find acceptable solutions; i.e., a solution that converges and matches reality.

[44] Schlieren, shadowgraph, and interferometer techniques have long been used in conventional wind tunnels for flow visualization of aerodynamic effects; e.g., shock wave and boundary layer effects. These methods are based on the optical property of a fluid that the speed of light varies with the density of the fluid through which it is passing.

7.1.2 Arc-Heated Facilities

Arc heated facilities, or arc-jet wind tunnels, heat air (or another gas) to very high temperatures to simulate atmospheric entry heating. These devices employ high-voltage electrical connectors around a cylindrical tube. As cold gas moves through the column it is heated by an electrical discharge and is expanded through a nozzle into a test chamber. The test chamber may be continuously evacuated by a vacuum system in order to maintain the desired test conditions.

Although arc-jet facilities have proven extremely valuable in terms of TPS materials selection and development, there are certain limitations. First, the chemical composition of the flow in flight is significantly different from that in an arc-jet facility. In flight, the gas is not dissociated or ionized upstream of the bow shock. Second, the upstream flow values are not constant; the flow approaching the test article is not necessarily uniform (flow velocity vectors may be divergent rather than parallel). Third, the Mach number and the Reynolds number are usually much lower than in flight making it difficult to simulate compressibility and viscous effects. Thus, the arc jet is an important tool for simulating high-enthalpy flows, but arc-jet facilities are poor wind tunnels in the classical sense. Performance values for NASA Ames Research Center's arc-jet test facilities are given in table 17.

TABLE 17. NASA AMES RESEARCH CENTER ARC-JET TEST FACILITIES

Facility	Test Gas	Mach Number	Test Article Size, cm	Bulk Enthalpy, MJ/kg	Surface Pressure, atm	Convective Heat Flux, W/cm^2
Aerodynamic Heating Facility (AHF), 20 MW	Air, Nitrogen	4 – 12	20 dia 66 x 66	12 – 33	0.001 – 0.125	0.06 – 255
Turbulent Flow Duct, 12 MW	Air, Nitrogen	3.5	20 x 51	3.5 – 9.3	0.02 – 0.15	2 – 68
Panel Test Facility (PTF) 20 MW	Air	5.5	35 x 35	4.6 – 33	0.0005 – 0.05	0.6 – 85
Interaction Heating Facility (IHF), 75 MW	Air	5.5 – 7.5	61 x 61 46 dia	7 – 46	0.0001 – 1.2	0.6 – 749[a]

[a] The Interaction Heating Facility can produce a radiative heat flux of 23 W/cm^2.

7.1.3 Shock Tunnel

A simple shock tube consists of two chambers containing gases at different pressures separated by a diaphragm. When the diaphragm is ruptured the high-pressure gas flows into the low-pressure chamber. If the pressure difference across the diaphragm is large enough, a supersonic flow and a travelling shock wave are created. Shock tubes provide valuable experimental data for high-speed gas flows but are limited by chamber pressures (or pressure ratio) and by chamber length. The consequences are very short run times and an inability to achieve hypersonic Mach numbers.

A shock tunnel overcomes some of the limitations of the shock tube by attaching a second diaphragm and a hypersonic nozzle at the end of the low-pressure chamber. When the second diaphragm is ruptured, gas flows through the nozzle and into a test section and then into a vacuum reservoir. It is possible to simulate dissociation, ionization and free molecular flow effects in a shock tunnel. Performance values for typical shock tunnels are:

Velocity	2 to 5 km/s
Reynolds number	3.3×10^4 to 3.3×10^8 per m
Mach number	6 to 24
Simulated Altitude	10 to 60 km
Run time	4 to 20 millisec

Note that run times are very short and flow velocities are much lower than Earth-entry velocities for Mars sample return missions. These values are for the U.S. Defense Department's Calspan shock-tunnel facility in Buffalo, NY (NASA Headquarters, 2006).

7.1.4 Ballistic Range Facilities

Ballistic range facilities are used to obtain static and dynamic aerodynamic data. In a ballistic range a model is fired from a gun into a test gas (usually a quiescent gas). Test data are measured and recorded at stations along the model flightpath. Test Reynolds numbers can be varied by increasing or decreasing the ambient pressure of the test gas. Ballistic range testing is discussed in an AGARD document by Canning et al. (1970) and in an AIAA paper by Strawa et al. (1988).

NASA Ames Research Center's Hypervelocity Free Flight Facility has been used extensively for testing a variety of entry vehicle configurations. The facility is capable of launching a 22 mm dia. model at a speed of 8 km/s. See Chapman and Yates (1998) for a description of the use of a ballistic range in the design of a planetary probe.

7.2 Flight Test

7.2.1 General Considerations

Ultimately in the design of an entry system, particularly one utilizing new materials or a new configuration, there is little substitute for an actual flight test.[45] This is the case if there are aspects of the flow field or configuration that are outside of the experimental database. Depending on the design aspect of specific interest (and of course the cost) different launch vehicles, flight velocities, or trajectories may be selected.

---

[45] The X-15 experimental aircraft achieved a maximum Mach number of 6.7 and a maximum altitude of 103.6 km in the 1960s. Three rocket-propelled, air-launched X-15s were built. They were launched from a B-52 aircraft and flew 199 flights from 1959 to 1969.

Note that in the National Aerospace Plane (NASP) program of the 1980s there was sufficient confidence in the CFD codes that experimental validation was considered unnecessary. However, it was later realized that flight testing was absolutely essential to "calibrate" and validate the computational models for boundary layer transition, engine inlet pressure distribution, etc. By that time, it was difficult to incorporate flight testing; this may have contributed to the subsequent cancellation of the program.

A simple sounding rocket can reach Mach 8 ($\approx$ 2.6 km/s) with relative ease. This allows meaningful aerodynamic experiments to be carried out at hypersonic speeds. For higher entry velocities, where aerodynamic heating becomes more significant, larger and more costly ballistic missiles may be used to achieve entry velocities on the order of 5 km/s. Most ICBMs, and many U.S. launch vehicles, fall into this category and have been extensively flight tested by the U.S. military. Boundary layer transition data were obtained on ICBM re-entry vehicle flight flights in the 1960s. High entry velocities can also be achieved using the "pile-driver" concept. The lower stages of the vehicle are used to gain altitude; then the final stage is pointed downward and its rocket engine ignited. The "pile-driver" method was used successfully in the NASA Ames Planetary Atmosphere Entry Technology (PAET) experiment—a four-stage Scout rocket launched from Wallops Island achieved an entry velocity of 7 km/s.

For testing above 7 km/s (roughly equivalent to re-entry from Earth orbit), the vehicle may require even greater "pile-driver" thrust. The NASA-proposed Aeroassist Flight Experiment (AFE) of the late 1980s was to be deployed from the Shuttle Orbiter and use a solid-rocket motor to thrust the vehicle into the atmosphere at approximately 12 km/s. Other experiments in this velocity range were done for U.S. and proposed Soviet lunar return capsules (Apollo and Soyuz). The cost of a flight experiment usually scales directly with entry velocity. For typical flight test entry velocities see table 18.

TABLE 18. TYPICAL FLIGHT TEST ENTRY VELOCITIES

Vehicle/System	Example	Velocity, V_E, km/s
Sounding Rocket	Terrier/Black Brant[a]	2.5
Sounding Rocket, Military ICBM	Scout, Minuteman	5
Shuttle Orbiter and "pile driver"	STS Orbiter, PAET	7+
Blunt ablator and "pile driver"	Apollo CM (unmanned)	9
Combination "pile driver"	Project Fire[b]	11.6

[a] See Jane's Space Directory (Baker, 2005).

[b] Project Fire was a NASA flight experiment with a blunt entry vehicle that achieved a velocity of 11.6 km/s during re-entry (Dingeldein, 1965). An Atlas launch vehicle was used with and Antares rocket motor fired at the end of the coast phase.

7.2.2 Impact and Recovery

Impact and recovery must be considered for every SRV experiment. Recovery phase design may have a significant bearing on how data is collected and analyzed and how accurately a probe or entry vehicle reaches its intended target. Significant issues affecting recovery are whether or not a parachute is used, effect of cross winds, atmospheric dispersions along the flightpath, and whether there is active guidance during any phase of the flight to correct accumulating errors.

For some flight projects the vehicle may reach the target area at supersonic speed. In this case all data must be transmitted and recorded before impact since the vehicle may be destroyed on impact. Sounding rockets often rely on parachute systems for terminal descent and a beacon for the recovery of payload or experiment capsules. The parachute also makes it easier to track, locate, and recover the data packages.

Parachute deceleration may be desirable for SRV capsule recovery. A parachute, even a small one, has the following advantages: easier tracking; more certainty of quickly finding the entry capsule; and a large decrease in g-loads at impact. The disadvantages include: increased system complexity associated with parachute deployment; a slight increase in the probe mass; and perhaps a lower static stability margin during hypersonic flight. Many of these questions and concerns can be resolved with sounding rocket flight tests.

8. DESIGN PROCESS SUMMARY

NASA's Stardust sample return mission returned safely to Earth when the capsule carrying cometary and interstellar particles successfully touched down at 2:10 a.m. Pacific time in the desert salt flats of the U.S. Air Force Utah Test and Training Range.

Ten years of planning and seven years of flight operations were realized early this morning when we successfully picked up our return capsule off of the desert floor in Utah.

The Stardust spacecraft released its sample return capsule at 9:57 p.m. Pacific time last night. The capsule entered the atmosphere four hours later. The drogue and main parachutes deployed at 2:00 and 2:05 a.m. Pacific time, respectively. The sample return capsule's science canister contained comet and interstellar dust. NASA's Stardust spacecraft traveled 2.88 billion miles during its seven-year round-trip odyssey.

NASA JPL Announcement, January 15, 2006

8.1   Typical Sample Return Mission (Stardust)

Stardust is a NASA comet rendezvous and sample return mission. The Stardust sample return capsule successfully returned to Earth in January 2006. The spacecraft captured particles from the tail of a comet. Two parachutes were employed during entry; a small one at 35.8 km altitude and the main chute at 3.05 km altitude. The Stardust sample return capsule is a blunt entry body (60-deg sphere-cone) with a PICA heatshield and SLA-561 ablation materials to protect the interior from entry heating. The sample return capsule entered the atmosphere at a speed of 12.5 km/s (relative to the atmosphere) and a flightpath angle of –8.4 deg. (The Stardust SRV is called a sample return capsule.) This was the fastest Earth entry ever for an artificial object.

Stardust is the first U.S. mission designed to return samples from another body since the Apollo missions to the moon; it is the first U.S. mission designed to obtain samples robotically in deep space and return them to Earth. Data for the successful Stardust mission are presented below to show typical values that could be encountered in a Mars sample return mission because the necessary entry velocity may be comparable (Jet Propulsion Laboratory Stardust Mission, 2006):

Launch Vehicle: Delta II (3 stages plus 4 solid rocket boosters attached to first stage)

Launch Date: February 1999 from Cape Canaveral

Distance of travel (entire mission): 4.63 billion km

Duration of Mission: 83 months in space

Power: Solar panels (6.6 m^2)

Earth entry velocity (inertial): 12.8 km/s

Landing site: U.S. Air Force Utah Test and Training Range

Size:

    Spacecraft: 1.7 m high x 0.66 m wide x 0.66 m deep
    Sample return capsule: 0.8 m dia x 0.5 m high

Mass:

Spacecraft	254 kg
Sample return capsule	46 kg
Fuel	85 kg
Total	385 kg

Program Cost: $168.4 million (not including launch vehicle)

The Stardust Earth-entry trajectory was simulated using the NASA TRAJ trajectory simulation code. Body geometry and initial conditions are given in table 19. Results of the trajectory calculation are shown in table 20. Trajectory variables (velocity, deceleration, heat flux and total heat load) are shown as functions of altitude and/or time from entry in figures 30 through 34. Table 20 shows that peak heat flux occurs prior to peak deceleration and peak dynamic pressure. Table 20 also shows that less than 10 percent of the total heat load is due to radiative heating. Figures 31 and 32 show the altitude of maximum deceleration and the altitude of maximum total (convective plus radiative) stagnation point heat flux, respectively.

Note the differences between inertial and relative velocities and between inertial and relative flightpath angles in table 19. The inertial values are the variables used in the equations of motion. The relative values are used to determine aerodynamic forces and moments and heat fluxes.

TABLE 19. STARDUST TRAJECTORY SIMULATION INPUT DATA

Variable	Value
Vehicle mass, m	41.5 kg
Cone half-angle, θ_C	60 deg
Nose radius, r_N	0.229 m
Base radius, r_B	0.406 m
Corner radius, r_C	0.019 m
Surface area, S_W	0.619 m^2
Ballistic Parameter, BP[a]	52.6 kg/m^2
Conditions at Entry Interface	
Altitude at entry, y_E	135 km
Angle of attack at entry, α	0 deg
Inertial velocity at entry	12.8 km/s
Relative velocity at entry	12.456 km/s
Inertial entry angle	−8.20 deg
Relative entry angle	−8.428 deg

[a] For altitudes greater than 37.5 km

64

TABLE 20. STARDUST TRAJECTORY SIMULATION OUTPUT DATA
(NASA TRAJ CODE, 1999)

a. Maximum values and corresponding trajectory points

Variable	Maximum Value	Time, s	Altitude, km	Velocity, km/s
Convective Heat Flux	605.4 W/cm^2	49.8	64.1	10.9
Radiant Heat Flux	102.9 W/cm^2	48.3	65.5	11.2
Total Heat Flux	705.1 W/cm^2	49.0	64.8	11.1
Stagnation Pressure	33,300 Pa	60.0	56.0	8.1
Dynamic Pressure	16,600 Pa	60.0	56.0	8.1
Deceleration Magnitude	315.7 m/s^2	60.0	56.0	8.1

b. Heat load at stagnation point

Heat Load	Value
Convective heat load, Q_{CONV}	18,290 J/cm^2
Radiative heat load, Q_{RAD}	1270 J/cm^2
Total heat load, Q_{TOTAL}	19,560 J/cm^2

Notes:
1976 U. S. Standard Atmosphere
Gravitational model includes J_2 harmonic
Destination planet is rotating Earth

8.2   Design Process for Sample Return Missions

The analytical, computational and experimental methods described in this document can be used in the design of an entry capsule for a Mars sample return mission. The process has several stages; some stages may require iteration.

8.2.1   Mission Design Requirements

The process can begin with maximum or minimum payload weight requirements. The Earth entry velocity is determined by the mission profile. For a return from Mars the entry velocity could be as high as 15 km/s. Mission requirements may limit the temperature in the capsule interior in order to protect the sample. The sample must survive landing impact.

8.2.2   Initial Probe Design

An overall conceptual design is developed based on initial assumptions for the fully deployed spacecraft along with SRV mass and stability considerations. The ballistic parameter is estimated from size, weight and estimated drag coefficient. This value of ballistic parameter is used to calculate atmospheric trajectories and perform parametric analyses of SRV variables.

The key variables in vehicle-trajectory optimization are entry velocity, entry angle and ballistic parameter for nonlifting, ballistic entry vehicles. The effect of varying the entry angle and ballistic parameter are discussed in section 4.

It is often possible to reduce maximum convective heat flux (and TPS weight) by reducing ballistic parameter and increasing flightpath angle at entry (more shallow entry). However, the total heat load may increase (requiring a thicker TPS) due to longer flight time.

Initial estimates of convective and radiative heat fluxes can be obtained from the analytical expressions given in Section 4 or from trajectory simulation codes. Note that estimates of radiative heat flux are less accurate than estimates of convective heat flux and total heat flux estimates are only approximate. With these heat flux estimates, an approximate TPS thickness can be determined and an approximate vehicle mass can be calculated.

What ultimately drives the selection of a TPS material and its thickness distribution is bondline, or attachment, temperature and total heat load. Consideration of total heat load is extremely critical when samples are very sensitive to temperature.

At later stages, more detailed flow field calculations are made to improve the convective and radiative heat flux predictions. This will allow more accurate determination of the TPS thickness distribution.

8.2.3   Terminal Descent and Recovery

One of the most important considerations in entry vehicle design is whether a parachute system should be included (see sec. 5). The advantages of parachute recovery are: a) entry vehicle tracking at high altitude is much easier; b) the probe or canister has a much lower ground impact velocity; and c) recovery operations are much easier (the probe is easier to find). Disadvantages may include: a) slight increase in mass; b) increased mechanical complexity with the separation of the probe from the aft-cover and parachute; and c) slight decrease in static stability margin.

A small parachute may be able to adequately decelerate the entry vehicle. Even though impact-area drift may increase significantly, ground radar should be able to track the probe and parachute and ensure efficient recovery operations.

8.3 Systems Engineering and Life Cycle Considerations

Systems engineering is the art and science of developing an operable system capable of meeting mission requirements within imposed constraints including (but not restricted to) mass, cost and schedule.

Griffin and French, 2004

Sample return vehicle design is just one element of the spacecraft mission life cycle process. Certainly SRV design and TPS material selection are critical elements and several iterations may be required to satisfy mission requirements and cost and schedule constraints. It is important to note some of the many other factors that are required for a successful spacecraft mission. NASA requires well-defined technical workflow and management reviews. These items are combined in NASA's systems engineering and technical integration process. NASA and many other organizations follow similar processes developed over many decades. NASA requires all significant projects to travel along this path from initial concept to mission completion (Shishko, 1995).

Project management and program control are part of the NASA project cycle. NASA programs follow a sequence that consists of distinct phases separated by "control gates." Failure to pass a control-gate review can result in project termination (see table 21).

TABLE 21. PROJECT LIFE CYCLE PHASE DESCRIPTIONS AND CONTROL GATES

Phase	Description	Control Gate
"Zero"	Advanced studies, mission feasibility	Mission Concept Review
A	Preliminary analysis, mission definition,	Mission Definition Review
B	System definition, preliminary design	Nonadvocate Review Preliminary Design Review
C	Design	Critical Design Review
D	Development, fabrication and integration, assembly, test	Operational Readiness Review Flight Readiness Review
E	Operations	—

Final consideration must be given to the many systems, subsystems, and support systems that must be factored into the project plan and schedule; these include:

Launch vehicle, environmental control

Ground support, data processing

Computers, hardware and software

Payload, lander and SRV

Attitude and thermal control

Electronics, power, batteries, propulsion

Communications, uplink and downlink

Vehicle and component structure

Operations support, tracking, command and data handling

The spacecraft must be maintained on optimal or near-optimal trajectories from launch to orbit, interplanetary, descent, landing, ascent, return, and Earth entry. All of this must be accomplished without compromising mission objectives.

8.4   Space Environment

There are many additional concerns in the design of a spacecraft mission (Griffin and French, 2004 and Fortescue et al., 2003). A few of these related to the space environment are listed below:

Vacuum – The near vacuum conditions in space lead to significant outgassing of materials. This outgassing can in turn result in changes in the electrical and mechanical properties of spacecraft materials.

Plasma effects – The effect of dissociation and ionization of gasses around the spacecraft and the SRV must be considered (Van Allen radiation belt effects, bombardment of the spacecraft by electrons, electromagnetic interference, and communications blackout during Earth entry).

Earth's magnetic field – The primary effect of variations in the Earth's magnetic field is on the spacecraft attitude control system.

Other radiation effects – Solar flares, x-rays, and ultra-violet radiation can affect electronic systems.

Planetary environment – Wind blown dust on the Martian surface as well as temperature extremes can affect the lander vehicle and the sample collection process. (Temperature extremes are also encountered in the interplanetary trajectories.)

Meteoroids and orbital debris – Meteoroids and orbital debris can cause damage to the spacecraft and the SRV.

Microgravity and weightlessness – Microgravity and weightlessness can affect spacecraft components and the flow of fluids.

REFERENCES

Allen, H. Julian: Motion of a ballistic missile angularly misaligned with the flight path upon entering the atmosphere and its effect upon aerodynamic heating, aerodynamic loads, and miss distance. NACA TN-4048, 1957.

Allen, H. Julian; and Eggers, Jr., A. J.: A study of the motion and aerodynamic heating of ballistic missiles entering the Earth's atmosphere at high supersonic speeds. NACA-TR-1381 (supersedes NACA-TN-4047), 1958.

Allen, Jr., Gary A.; Wright, Michael J.; and Gage, Peter: The trajectory code (TRAJ): Reference manual and user's guide. NASA/TM–2005-212847, March 2005.

Allen, W. H.: Dictionary of technical terms for aerospace use. NASA SP-7, 1965.

American Institute of Aeronautics and Astronautics: Bibliographic database of AIAA publications and technical papers, 1992 to present.
http://www.aiaa.org (Accessed June 2006)

Ames Research Staff, Ames Aeronautical Laboratory: Equations, tables, and charts for compressible flow. National Advisory Committee for Aeronautics. NACA-TR-1135, 1953.

Anderson, Jr., John D.: Computational fluid dynamics, the basics with applications. McGraw-Hill, 1995.

Anderson, Jr., John D.: Hypersonic and high temperature gasdynamics. McGraw-Hill, 1989.

Ashley, Holt: Engineering analysis of flight vehicles. Dover Reprint, 1992 (originally published by Addison-Wesley, 1974).

Avallone, E. A.; and Baumeister III, T. (editors): Marks' standard handbook for mechanical engineers (10th edition). McGraw-Hill, 1996.

Baker, David (ed.): Jane's space directory (21st edition). Jane's Information Group, 2005.

Bertin, John J.: Hypersonic aerothermodynamics (AIAA Education Series). American Institute of Aeronautics and Astronautics, 1994.

Blake, Bernard: Jane's weapon systems (19th Edition), 1988-89. Jane's Information Group, 1988.

Brauer, G. L.; Cornick, D. E.; Habeger, A.R.; Petersen, F.M.; and Stevenson, R.: Program to Optimize Simulated Trajectories (POST), 2 vols. NASA-CR-132689 and NASA-CR-132690, 1975.

Canning, T. N.; Seiff A.; and James, C.S. (eds.): Ballistic-range technology. AGARD-AG-138-70, Advisory Group for Aerospace Research and Development, Paris, France, 1970.

CFD-Online: Computational Fluid Dynamics Resources Online (an annotated list of CFD resources). http://www.cfd-online.com/Resources/ (Accessed June 2006)

Chapman, Dean R.: An analysis of the corridor and guidance requirements for supercircular entry into planetary atmospheres. NASA-TR-R-55, 1960.

Chapman, Dean R.: An approximate analytical method for studying entry into planetary atmospheres. NASA-TR-R-11, 1959.

Chapman, Dean R.; and Kapphahn, A. K.: Tables of Z functions for atmosphere entry analyses. NASA TR-R-106, 1961.

Chapman, Gary; and Yates, Leslie A.: Dynamics of planetary probes: design and testing issues. AIAA Paper 98-0797, 36th AIAA Aerospace Sciences Meeting, Reno, NV, 1998.

Chen, Y. -K.; and Milos, F. S.: Ablation and thermal response program for spacecraft heatshield analysis. AIAA Paper 98-0273, 36th AIAA Aerospace Sciences Meeting, Reno, NV, 1998.

Chen, Y. -K.; and Milos, F. S.: Fully Implicit Ablation and Thermal Analysis Program (FIAT). Proc. Fourth International Conference on Composites Engineering, College of Engineering, University of New Orleans, July 1997.

Delft Technical University, Netherlands: Aerodynamic flow calculator (calculator for compressible flow, shock wave, and Prandtl-Meyer expansion properties). http://www.aero.lr.tudelft.nl/~bert/flowcalc.html (Accessed June 2006)

Desai, P. B; Mitcheltree, R. A.; and Cheatwood, F. M.: Sample returns missions in the coming decade. IAF Paper-00-Q.2.0.4, 5th International Astronautical Congress, Rio de Janeiro, Brazil, 2000.

Devenport, William J.: Compressible Aerodynamics Calculator (compressible flow and shock wave properties for perfect gases). Dept. of Aerospace and Ocean Engineering, Virginia Polytechnic Institute and State University. http://www.aoe.vt.edu/aoe3114/calc.html (Accessed June 2006)

Dingeldein, R. C.: Flight measurements of reentry heating at hyperbolic velocity (Project Fire). NASA TM X-1053, 1965.

Eggers, A. J., Jr.; Allen, H. Julian; and Neice, S. E.: A comparative analysis of the performance of long-range hypervelocity vehicles. NACA-TR-1382, 1958.

Etkin, Bernard: Dynamics of atmospheric flight. Wiley, 1972.

Fay, J. A.; and Riddell, F. R.: Theory of stagnation point heat transfer in dissociated air. J. of the Aeronautical Sciences, vol. 25, no. 2, 1958, pp. 73-85.

Fehlberg, E.: Low-order classical Runge-Kutta formulas with stepsize control and their application to some heat transfer problems. NASA TR-R-315, July 1969.

Forsythe, G. E.; Malcolm, M. A.; and Moler, C. B.: Computer methods for mathematical computations. Prentice-Hall, 1977.

Fortescue, P.; Stark, J.; and Swinerd, G. (eds.): Spacecraft Systems Engineering (3rd edition). Wiley, 2003.

Glover, Jr., D. R.: Definition of Technical Terms for Aerospace Use. http://roland.lerc.nasa.gov/~dglover/dictionary//content.html (Accessed June 2006)

Gordon, S.; and McBride, B. J.: Computer Program for Calculation of Complex Chemical Equilibrium Compositions and Applications, I. Analysis. NASA RP-1311, 1994.

Griffin, Michael D.; and French, James R.: Space Vehicle Design (2nd edition). AIAA Education Series, American Institute of Aeronautics and Astronautics, 2004.

Hansen, C. Frederick; and Heims, Steven P.: A review of the thermodynamic, transport, and chemical reaction rate properties of high-temperature air. NACA TN-4359, 1958.

Hoerner, Sighard F.: Fluid-dynamic drag: practical information on aerodynamic drag and hydrodynamic resistance. Sighard F. Hoerner Fluid Dynamics, 1965.

Hoerner, Sighard F.; and Borst, Henry V.: Fluid-dynamic lift: practical information on aerodynamic and hydrodynamic lift (2nd edition). Hoerner Fluid Dynamics, 1985.

Howe, John T.: Hypervelocity atmospheric flight: real gas flow fields. NASA RP-1249, 1990.

Hughes, William F.; and Brighton, John A.: Schaum's outline of theory and problems of fluid dynamics (3rd edition). McGraw Hill, 1999.

Jakob, Max: Heat transfer (Vol. 1). Wiley, 1949.

Jet Propulsion Laboratory, California Institute of Technology: Mars Exploration Website. http://mars.jpl.nasa.gov (Accessed June 2006)

Jet Propulsion Laboratory, California Institute of Technology: Stardust Mission Website. http://stardust.jpl.nasa.gov (Accessed June 2006)

Karamcheti, K.: Principles of ideal-fluid aerodynamics. Krieger Publishing, 1980 (originally published by Wiley, 1966).

Katz, Joseph; and Plotkin, Allen: Low-speed aerodynamics: from wing theory to panel methods. McGraw Hill, 1991.

Kolodziej, Paul: Aerothermal Performance Constraints for Hypervelocity Small Radius Unswept Leading Edges and Nosetips. NASA TM-112204, 1997.

Kolodziej, Paul; Bowles, Jeffery V.; and Roberts, Cathy: Optimizing hypersonic sharp body concepts from a thermal protection perspective. AIAA Paper 98-1610, 8th AIAA International Space Planes and Hypersonics Conference, Atlanta, GA, 1998.

Kopal, Z.: Tables of supersonic flow around cones. Technical Report 1, Massachusetts Institute of Technology, 1947.

Liepmann, H. W.; and Roshko, A.: Elements of gasdynamics. Wiley, 1957.

Lukasiewicz, J.: Experimental methods of hypersonics. M. Decker, 1973.

McCormick, Barnes W.: Aerodynamics, aeronautics, and flight mechanics (2nd edition). Wiley, 1995.

Mitcheltree, R. A.; Kellas, S.; Dorsey, J. T.; Desai, P. N.; and Martin, C. J.: A Passive Earth-Entry Capsule for Mars Sample Return. AIAA Paper No. 98-2851, 7th AIAA/ASME Joint Thermophysics and Heat Transfer Conference, Albuquerque, NM, 1998.

NASA Glenn Research Center: Learning Technologies Project Website. Navier-Stokes equations and other subjects in aeronautics for students and educators. http://www.grc.nasa.gov/WWW/K-12/airplane/nseqs.html (Accessed June 2006)

NASA Headquarters: Major Facility Inventory (data on major U.S. Government test facilities). Contact: Cheryl DiLustro, NASA Headquarters, MS 5C63, 300 E St. SW, Washington DC 20546-0001, sherry.dilustro@nasa.gov, 1-202-358-1124. User ID and Password required. http://facility.public.hq.nasa.gov (Accessed June 2006)

NASA Scientific and Technical Information (STI): NASA aeronautics and space database. Bibliographic database provides access to 3.9 million citations from 1915 to the present. Many full-text downloads available. For NASA internal user only http://sti.nasa.gov (Accessed June 2006)

NASA Technical Reports Server (NTRS): NASA documents available to the general public. http://ntrs.nasa.gov/search.jsp (Accessed June 2006)

National Oceanic and Atmospheric Administration: U.S. Standard Atmosphere, 1976. U.S. Government Printing Office, 1976.

Open Channel Foundation: NASA COSMIC Collection. Open Channel Software, 1807 W. Sunnyside Avenue, Suite 301, Chicago, IL 60640, 1-773-334-8177 http://openchannelsoftware.org/cosmic (Accessed June 2006)

Pope, Alan; and Goin, Kenneth L.: High-Speed Wind Tunnel Testing. Krieger Publishing, 1978 (originally published by Wiley, 1965).

Portree, David S. F.: NASA's origins and the dawn of the space age. NASA Monographs in Aerospace History #10, 1998.

Regan, Frank J.: Re-entry vehicle dynamics. American Institute of Aeronautics and Astronautics, 1984.

Regan, Frank J.; and Anandakrishnan, Satya M.: Dynamics of atmospheric re-entry. American Institute of Aeronautics and Astronautics, 1993.

Rohsenow, W. M.; Hartnett, J. P.; and Ganic, E. N. (eds.): Handbook of Heat Transfer Fundamentals (2nd edition). McGraw-Hill, 1985.

Scarborough, James B.: Numerical mathematics analysis (6th edition). Johns Hopkins Press, 1966.

Schlichting, Hermann; and Gersten, Klaus: Boundary-layer theory (8th edition). Springer-Verlag, 2000.

Shishko, Robert: NASA Systems Engineering Handbook. NASA SP-6105, 1995.

Sims, J. L.: Tables for supersonic flow around right circular cones at zero angle of attack. NASA SP-3004, 1964.

Sims, J. L.: Tables for supersonic flow around right circular cones at small angle of attack. NASA SP-3007, 1964.

Sommer, Simon C.; and Tobak, Murray: Study of the oscillatory motion of manned vehicles entering the Earth's atmosphere. NASA Memo 3-2-59A, 1959.

Spreiter, John R.: Transonic aerodynamics (Chapter 1), Progress in Astronautics and Aeronautics, Vol. 81, David Nixon (ed.), American Institute of Aeronautics and Astronautics, 1982.

Squire, Thomas H.: TPSX Material Properties Database. Web Ed., Version 4.
Contact: Thomas H. Squire, NASA Ames Research Center, MS 234-1, Moffett Field, CA 94035-1000, Thomas.H.Squire@nasa.gov, 1-650-604-1113
http://tpsx.arc.nasa.gov (Accessed June 2006)

Strawa, A. W.; Chapman, G. T.; Arnold, J. O.; and Canning, T. N.: The ballistic range and aerothermodynamic testing. AIAA Paper 88-2015, 15th Aerodynamic Testing Conference, San Diego, CA, 1988.

Tauber, M. E.: A brief review of some mechanisms causing boundary layer transition at high speeds. NASA-TM-102834, 1990.

Tauber, M. E.: A review of high-speed, convective, heat-transfer computation methods. NASA TP-2914, 1989.

Tauber, M. E.; and Sutton, K.: Stagnation-point radiative heating relations for Earth and Mars entries. J. of Spacecraft and Rockets, vol. 28, no. 1, 1991, pp. 40-42.

Tauber, M.: Atmospheric entry (Lecture notes for AA 213). Stanford University Bookstore, 1998.

Tobak, Murray; and Allen, H. Julian: Dynamic stability of vehicles traversing ascending or descending paths through the atmosphere. NACA TN-4275, 1958.

Van Dyke, M. D.: The combined supersonic-hypersonic similarity rule. J. of the Aeronautical Sciences, vol. 18, no. 7, July 1951, pp. 499-500.

Wercinski, P. F.: Mars Sample Return: A Direct and Minimum-Risk Design. J. of Spacecraft and Rockets, vol. 33, no. 3, 1996, pp. 381-385.

White, Frank M.: Viscous fluid flow (3rd edition). McGraw-Hill, 2005.

Wurster, K. E.; Riley, C. J.; and Zoby, E. V.: Engineering Aerothermal Analysis for X-34 Thermal Protection System Design. J. of Spacecraft and Rockets, vol. 36, no. 2, 1999, pp. 216-228.

Zwillinger, Daniel (ed.): CRC Standard Mathematical Tables and Formulae (31st edition). Chapman & Hall CRC Press, 2002.

APPENDIX

Acronyms and Abbreviations

AEDC	Arnold Engineering Development Center
AFE	Aeroassist Flight Experiment
AFRSI	Advanced flexible reusable surface insulation
AGARD	Advisory Group for Aerospace Research and Development
AHF	Aerodynamic Heating Facility
AIAA	American Institute of Aeronautics and Astronautics
ANSI	American National Standards Institute
AOTV	Aero-Assist Orbital Transfer Vehicle
ASTM	American Society for Testing and Materials
AVCOAT	Trade name for ablation material developed by AVCO Corp.
CFD	Computational fluid dynamics
CM	Command Module (Apollo)
DOF	Degrees of freedom
DSMC	Direct simulation Monte Carlo
FIAT	Fully implicit ablation and thermal response program
HAVOC	Hypersonic vehicle optimization code
ICBM	Intercontinental ballistic missile
IEEE	Institute of Electrical and Electronic Engineers
IHF	Interaction Heating Facility
JPL	Jet Propulsion Laboratory
LCA	Lightweight ceramic ablator
LEO	Low Earth orbit
NACA	National Advisory Committee for Aeronautics
NASA	National Aeronautics and Space Administration
NASP	National Aero-Space Plane
NIST	National Institute of Standards and Technology
OMS	Orbital maneuvering system
PAET	Planetary Atmosphere Entry Technology

PICA	Phenolic impregnated carbon ablator
POST	Program to optimize simulated trajectories
PTF	Panel Test Facility
RCC	Reinforced carbon-carbon composite
RCG	Reaction cured glass
SI	International System of Units
SLA	Superlight ablator
SRV	Sample return vehicle
STS	Space Transportation System
TABI	Tailorable advanced blanket insulation
TPS	Thermal protection system
TPSX	Ames database of TPS material properties
TRAJ	Trajectory code
TUFI	Toughened uni-piece fibrous insulation

Nomenclature

α	angle of attack
β	inverse scale height
χ	viscous interaction parameter
δ	boundary layer thickness
ε	radiation emissivity of material surface
ϕ	bank angle (measured from vertical)
γ	flightpath angle (measured from horizontal, positive up); or isentropic exponent (ratio of specific heats, c_p/c_v)
λ	mean free path, or Earth latitude
μ	dynamic, or absolute viscosity
θ	angle between tangent to body surface and freestream direction; or angle between body axis and body surface
ρ	mass density
σ	Stefan-Boltzmann constant, $5.670\,400 \times 10^{-8}$ W/(m^2-K^4)
ψ	heading angle (measured from entry plane)
ν	kinematic viscosity
τ	fineness ratio

76

ω	frequency of oscillation
A	axial force
BP	ballistic parameter
c	speed of sound
C	Chapman-Rubesin constant
CG	Center of mass (or center of gravity) location
CP	center of pressure location
C_A	axial force coefficient
C_D	drag coefficient
C_{D_0}	zero-lift drag coefficient
C_f	skin friction drag coefficient
C_L	lift coefficient
C_m	pitching moment coefficient
C_{m_α}	pitching moment curve slope
C_N	normal force coefficient
C_{N_α}	normal force curve slope
C_p	pressure coefficient
c_p	specific heat at constant pressure
c_v	specific heat at constant volume
D	drag force; or diameter
g	acceleration due to gravity
h	enthalpy per unit mass; or geopotential altitude
I	mass moment of inertia
J_2	zonal harmonic coefficient in a series representing the Earth's gravity field (primarily the effects of Earth oblateness)
K	arbitrary constant
k_1	dynamic stability parameter
k_2	static stability parameter
k	thermal conductivity
Kn	Knudsen number
L	lift force, or vehicle reference length
Le	Lewis number
m	mass

M	Mach number, or pitching moment
M	molecular weight
N	normal force
p	pressure
Pr	Prandtl number
q	dynamic pressure; or heat flux (quantity of heat per unit area and per unit time)
Q	total heat load
r	radius, or distance from Earth center to vehicle center of mass
R	gas constant, or Earth radius
R	Universal gas constant
Re	Reynolds number
Re_D	Reynolds number based on diameter D, $\rho \, V \, D \, / \, \mu$
Re_L	Reynolds number based on length L, $\rho \, V \, L \, / \, \mu$
Re_x	Reynolds number based on distance x, $\rho \, V \, x \, / \, \mu$
S	reference area, or surface area
St	Stanton number
T	temperature
t	time measured from entry
V	freestream or vehicle velocity magnitude, or modified viscous interaction parameter
W	weight
x	axial, or longitudinal coordinate
y	geometric altitude, or coordinate normal to body surface

Subscripts

0	sea level conditions
1, 2	conditions upstream and downstream of shock wave, respectively
AW	adiabatic wall conditions
B	conditions at base of body
C	cone, or corner between cone surface and base
CG	center of gravity
COND	conductive heat transfer
CONV	convective heat transfer

CP	center of pressure
D	diameter
e	boundary layer outer edge conditions
E	conditions at entry interface altitude
F	characteristic fluid temperature (usually mean or freestream)
FLIGHT	flight conditions
INCOMP	incompressible flow
L	reference length
LAM	laminar flow
M	molecular scale
MAX	maximum
N	conditions at nose of body
RAD	radiative heat transfer (from shock layer gases to vehicle surface)
RE-RAD	re-radiative heat transfer (from vehicle surface to space)
SPACE	radiation to space or radiation sink
t	stagnation or conditions at stagnation point
T	approximate location where boundary layer transition (from laminar flow to turbulent flow) occurs
TOTAL	convective plus radiative heat transfer, CONV plus RAD
TRANS	boundary layer transition
TURB	turbulent flow
W	wall conditions or body surface area
WIND TUNNEL	wind tunnel conditions and model scale
x	distance along surface flat plate
∞	freestream conditions

Units

atm	atmosphere
bar	10^{-5} Pa
Btu	British thermal unit
deg	degree
ft	foot

g   gram

h   hour

in   inch

J   joule

K   Kelvin

lb   pound mass

lbf   pound force

m   meter

mi   mile

min   minute (time)

mol   mole

N   newton

ºC   degree Celsius

ºF   degree Farenheit

ºR   degree Rankine

Pa   pascal

psf   pound per square foot

psi   pound per square inch

rad   radian

s   second

slug   unit of mass in foot-pound-second system

W   watt

Constants and Conversion Factors

1 deg = $(\pi/180)$ rad = 0.01745329 rad (plane angle)

π = 3.14159265358979 …

e = 2.718281828 … (base of natural logarithms)

1 Btu = 1055.056 J (International Table) = 1054.350 J (Thermochemical)

1 J = 1 N-m

1 W = 1 J/s

1 Pa = 1 N/m^2

Length

1 in = 2.54 cm

1 ft = 0.3048 m

1 mi = 5280 ft = 1609.344 m

1 nautical mile = 1852 m = 1.15077945 mi

Area

$1 \text{ in}^2 = 6.4516 \text{ cm}^2$

$1 \text{ ft}^2 = 0.092903 \text{ m}^2$

Volume

$1 \text{ in}^3 = 16.387064 \text{ cm}^3 = 0.016387 \text{ liter}$

$1 \text{ ft}^3 = 0.028317 \text{ m}^3 = 28.317 \text{ liter}$

Speed

1 mile per hour = 0.44704 m/s

1 ft/s = 0.3048 m/s = 0.0003048 km/s

Force

1 lbf = 4.448222 N

Pressure

$1 \text{ bar} = 100 \text{ kPa} = 100,000 \text{ N/m}^2$

$1 \text{ psi} = 1 \text{ lbf/in}^2 = 6894.7572 \text{ Pa} = 6894.7572 \text{ N/m}^2$

$1 \text{ psf} = 1 \text{ lbf/ft}^2 = 47.88026 \text{ Pa} = 47.88026 \text{ N/m}^2$

1 atm = 101,325 Pa = 760 mm Hg = 760 torr

$1 \text{ atm} = 1.01325 \text{ bar} = 14.696 \text{ lbf/in}^2$

Mass and Density

1 lb = 0.45359237 kg

1 slug = 14.59390 kg

$1 \text{ lb/ft}^3 = 16.018046 \text{ kg/m}^3 = 0.031081 \text{ slug/ft}^3$

Energy and Power

1 ft-lbf = 1.355818 J

1 ft-lbf/s = 1.355818 W

1 Btu/s = 1055.056 J/s = 1055.056 W

1 Btu/(ft^2-s) = 1.135653 x 10^{-4} W/m^2 (heat flux)

1 Btu/^{o}R = 1899.101 J/K (heat capacity)

1 Btu/(lb-^{o}F) = 4186.8 J/(kg-K) (specific heat capacity)

1 Btu/lb = 2.3218 x 103 N-m/kg (specific enthalpy)

Fluid Properties

1 lb/ft-s = 1.488164 Pa-s (viscosity)

1 Btu-in/(s-ft^2-^{o}F) = 519.2204 W/(m-K) (thermal conductivity)

1 Btu/(s-ft^2-^{o}F) = 20,441.75 W/(m^2-K) (heat transfer coefficient)

Temperature

T(K)   = 5/9 T(^{o}R)

T(K)   = T(^{o}C) + 273.15

T(K)   = (5/9) [T(^{o}F) + 459.67]

T(^{o}C) = (5/9) [T(^{o}F) – 32]

T(^{o}R) = T(^{o}F) + 459.67

1 ^{o}C   = 5/9 ^{o}F

International System of Units (SI) Prefixes

n	nano	10^{-9}
μ	micro	10^{-6}
m	milli	10^{-3}
c	centi	10^{-2}
k	kilo	10^{3}
M	mega	10^{6}
G	giga	10^{9}

82

Bibliography

American Society for Testing and Materials: Standard for use of the international system of units (SI): the modern metric system. IEEE/ASTM SI-10 (Replaces ASTM E 380 and ANSI/IEEE Std 268-1992). American Society for Testing and Materials, 1999.

Anderson, Jr., John D.: Fundamentals of aerodynamics (3rd edition). McGraw-Hill, 2001.

Anderson, Jr., John D.: Modern compressible flow: with historical perspective (3rd edition). McGraw-Hill, 2003.

Barter, N. J. (ed.): TRW Space Data (5th edition). TRW Inc., Space & Electronics Group, 1999.

Bertin, John J.; and Smith, Michael L.: Aerodynamics for engineers (3rd edition). Prentice Hall, 1998.

Binder, Raymond C.: Fluid Mechanics (5th edition). Prentice-Hall, 1973.

Bird, G. A.: Molecular Gas Dynamics and the Direct Simulation of Gas Flows. Oxford Engineering Science Series No. 42. Clarendon Press, 1994.

Blake, Bernard (ed.): Jane's weapon systems, 1988-1989 (19th Edition). Jane's Information Group, 1988.

Curtis, Howard D.: Orbital mechanics for engineering students. Elsevier, 2005.

Davies, Mark (editor): The standard handbook for aeronautical and astronautical engineers. McGraw-Hill, 2003.

Defence Science and Technology Laboratory: British Defence Science and Technology Laboratory Website (formerly a part of British Defence Evaluation and Research Agency). http://www.dstl.gov.uk/ (Accessed June 2006)

Defense Technical Information Center (DTIC): U.S. Department of Defense Scientific and Technical Information Website. http://www.dtic.mil (Accessed June 2006)

Eckert, E. R. G.; and Drake, Jr., Robert M.: Heat and mass transfer (2nd edition). McGraw-Hill, 1959.

Engineering Information Inc.: Ei Engineering Village 2 (bibliographic database with files from 1970, offering over five million summaries of journal articles, technical reports, and conference papers and proceedings in electronic form). http://www.ei.org (Accessed June 2006)

Escobal, P. R.: Methods of Orbit Determination. Krieger Publishing, 1976 (originally published by Wiley, 1965).

European Space Agency (ESA): Website of the European Space Agency. http://www.esa.int/esaCP/index.html (Accessed June 2006)

Fox, R. W.; and McDonald, A. T.: Introduction to fluid mechanics (4th edition). Wiley, 1992.

Gnoffo, Peter A: Planetary-Entry Gas Dynamics. Annual Review of Fluid Mechanics, vol. 31, Jan. 1999, pp. 459-494.

Hagen, K. D.: Heat transfer with applications. Prentice Hall, 1999.

Hale, Francis J.: Introduction to space flight. Prentice-Hall, 1994.

Hankey, Wilbur L.: Re-entry aerodynamics. AIAA Education Series, American Institute of Aeronautics and Astronautics, Washington, D.C., 1988.

Hayes, Wallace D.; and Probstein, Ronald F.: Hypersonic Inviscid Flow. Dover, 2004 (originally published by Academic Press as Hypersonic Flow Theory (vol. 1), 1966).

Incropera, Frank P.; and DeWitt, David P.: Fundamentals of heat and mass transfer (4th edition). Wiley, 1996.

Incropera, Frank P.: Introduction to heat transfer (3rd edition). Wiley, 1996.

Jet Propulsion Laboratory, California Institute of Technology: Basics of Spaceflight. http://www2.jpl.nasa.gov/basics/ (Accessed June 2006)

Kroo, Ilan: Applied Aerodynamics: A Digital Textbook. http://www.desktopaero.com/appliedaero (Accessed June 2006)

Kuethe, Arnold M.; and Chow, Chuen-Yen: Foundations of aerodynamics: bases of aerodynamic design (5th edition). Wiley, 1998.

Library of Congress: Library of Congress Website and Catalog. http://catalog.loc.gov (Accessed June 2006)

Liggett, James A.; and Caughey, David A.: Fluid mechanics: an interactive text (CD-ROM). ASCE Press, 1998.

Loh, W. H. T.: Re-entry and planetary entry physics and technology (vols. 2 and 3). Springer-Verlag, 1968.

Martin, John J.: Atmospheric reentry: an introduction to its science and engineering. Prentice-Hall, 1966.

McAdams, William H.: Heat transmission (3rd edition). McGraw-Hill, 1954.

McBride, Bonnie J.; Zehe, Michael J.; and Gordon, Sanford: NASA Glenn Coefficients for Calculating Thermodynamic Properties of Individual Species. NASA/TP—2002-211556, 2002.

Mechtly, E. A.: The international system of units: physical constants and conversion factors (2nd revision). NASA SP-7012, 1973.

Meyer, Rudolf X.: Elements of space technology. Academic Press, 1999.

Miele, Angelo: Flight mechanics. Addison-Wesley, 1962.

NASA Galaxie: Catalogs of NASA HQ and NASA field center libraries (for NASA internal use only). http://nasagalaxie.larc.nasa.gov (Accessed June 2006)

National Institute of Standards and Technology (NIST): International System of Units (SI). Includes online versions of NIST SP-330 and NIST SP-811. http://physics.nist.gov/cuu/Units/index.html (Accessed June 2006)

Office National d'Études et de Recherches Aérospatiales (ONERA): French National Aerospace Agency website. http://www.onera.fr/english.html (Accessed June 2006)

Park, Chul: Nonequilibrium hypersonic aerothermodynamics. Wiley, 1990.

Petrik Laboratories: PETRIK Library Scientific Conversion Chart. http://www.petrik.com/PUBLIC/library/conv/conversionsAE.htm (Accessed June 2006)

Rouse, Hunter: Elementary mechanics of fluids. Dover, 1978 (originally published by Wiley, 1946).

Streeter, Victor L.; Wylie, E. Benjamin; and Bedford, Keith W.: Fluid mechanics (9th edition). WCB/McGraw Hill, 1998.

Sutton, George P.: Rocket propulsion elements: an introduction to the engineering of rockets (7th edition). Wiley, 2001.

Taylor, Barry N. (ed.): The international system of units (SI). Translation of the 6th edition of the International Bureau of Weights and Measures publication "Le Systeme International d'Unities." National Institute of Standards and Technology, SP-330, 2001 (Supersedes NBS-SP-330). http://physics.nist.gov/Document/sp330.pdf (Accessed June 2006)

Taylor, Barry N.: Guide for use of the international system of units (SI). National Institute of Standards and Technology, NIST SP-811, 1995. http://physics.nist.gov/Document/sp811.pdf (Accessed June 2006)

Thompson, William T.: Introduction to Space Dynamics. Wiley, 1961.

Van Dyke, M. D.: An album of fluid motion. Parabolic Press, 1982.

Vincenti, Walter G.; and Kruger, Jr., Charles H.: Introduction to physical gas dynamics. Krieger Publishing, 1982 (originally published by Wiley, 1965).

Vinh, Nguyen X.: Optimal trajectories in atmospheric flight. Elsevier Scientific, 1981.

Von Karman Institute: Capsule aerothermodynamics. AGARD-R-808, Advisory Group for Aerospace Research and Development, 1995.

Wendt, John F.: Hypersonic aerothermodynamics (Lecture Series 1984-01). Von Karman Institute for Fluid Dynamics, February 1984.

White, Frank M.: Fluid Mechanics (5th edition). McGraw-Hill, 2003.

Wiesel, William E.: Spaceflight dynamics (2nd edition). McGraw-Hill, 1997.

Wikipedia: Encyclopedia Astronautica (a reference website on space travel). http://en.wikipedia.org/wiki/Encyclopedia_Astronautica (Accessed June 2006)

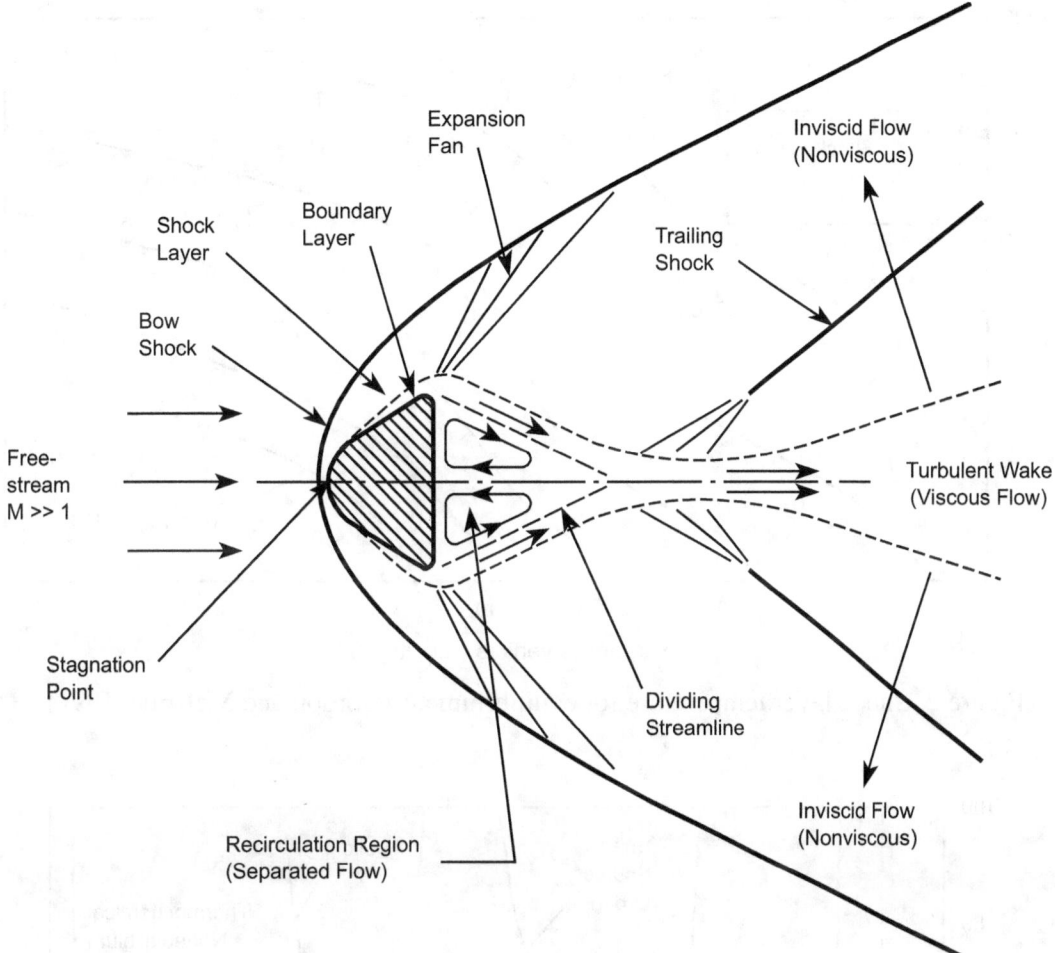

Figure 1. Characteristics of hypersonic flow around a blunt body.

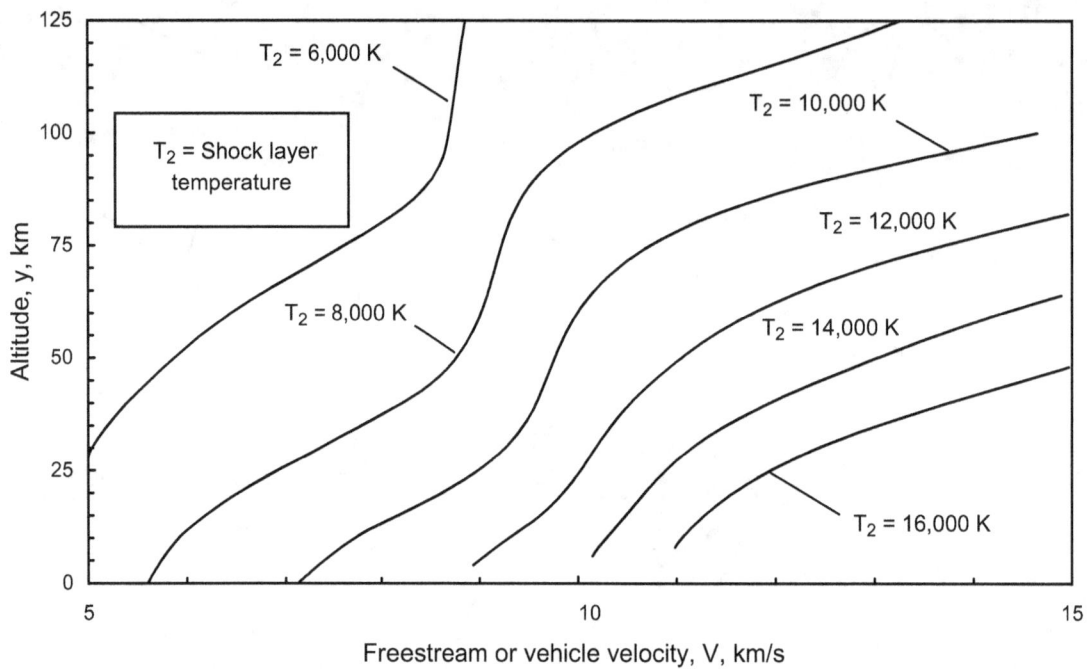

Figure 2. Shock layer temperature for equilibrium air (Gordon and McBride, 1994).

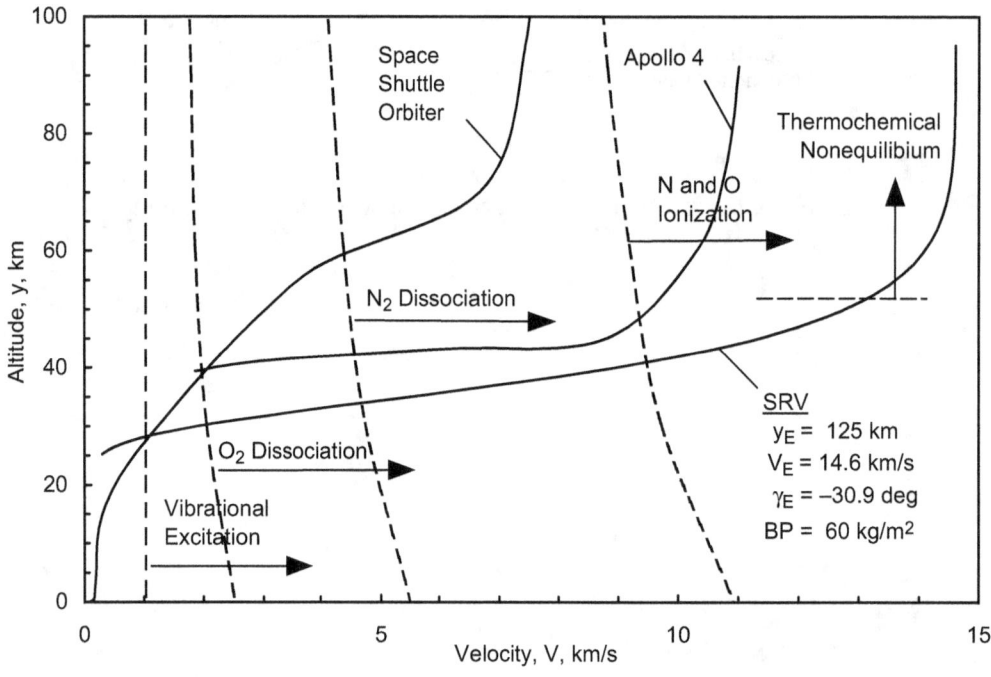

Figure 3. Vehicle flight profiles in Earth atmosphere (Hansen and Heims, 1958 and Howe, 1990).

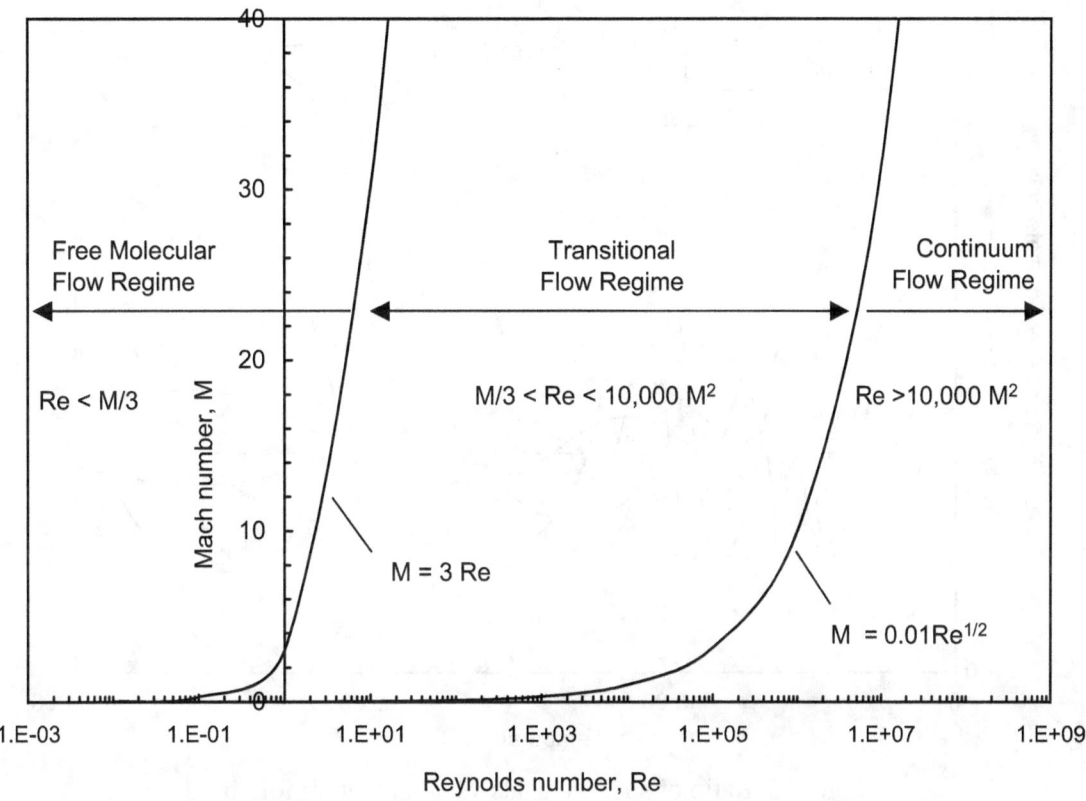

Figure 4. Flow regimes based on Mach number and Reynolds number.

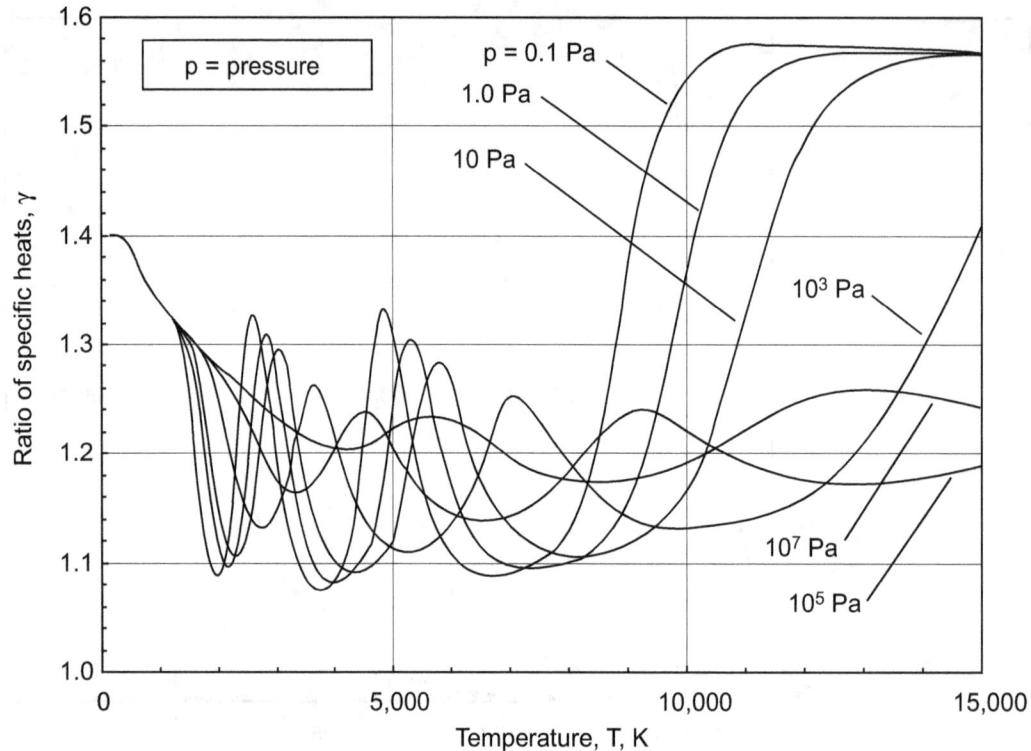

Figure 5. Ratio of specific heats vs. temperature for air.

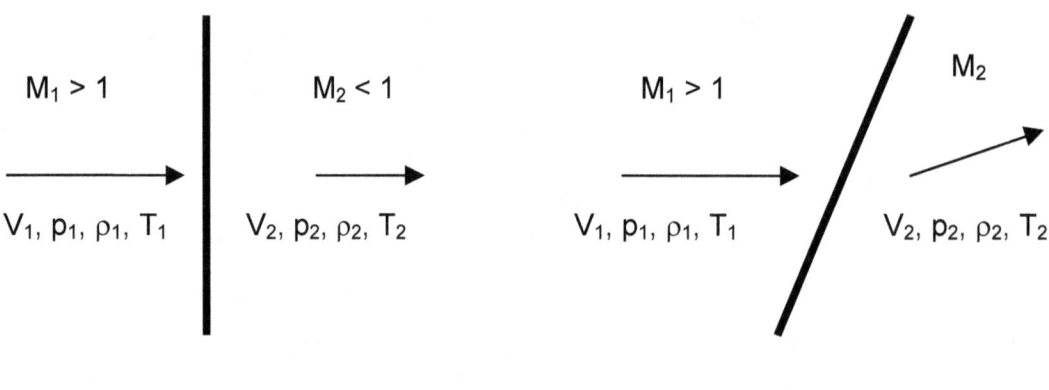

a. Normal shock wave   b. Oblique shock wave

Figure 6. Normal and oblique shock waves.

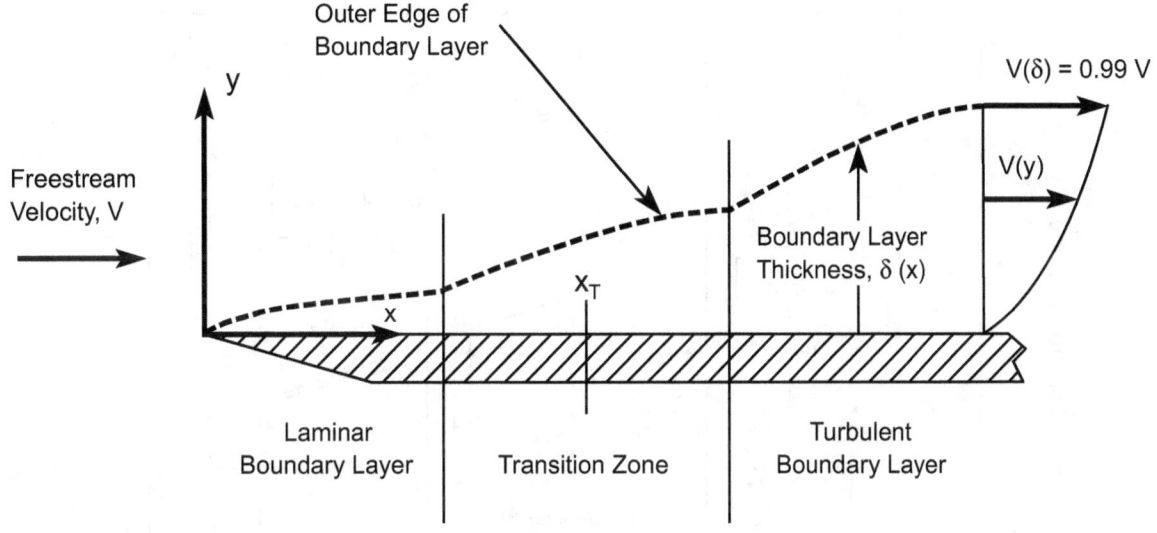

Notes:

x_T is assumed location of transition from laminar to turbulent flow

y-coordinate is greatly enlarged.

Figure 7. Boundary layer on a flat plate.

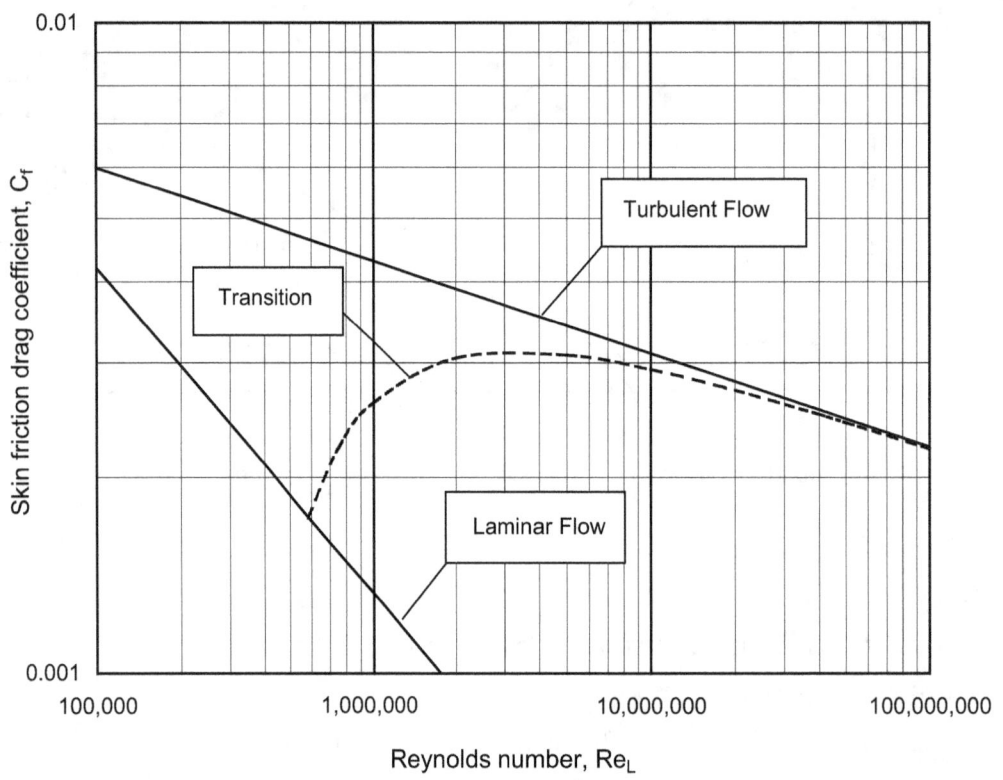

Figure 8. Skin friction drag coefficient vs. Reynolds number for flat plate in incompressible flow.

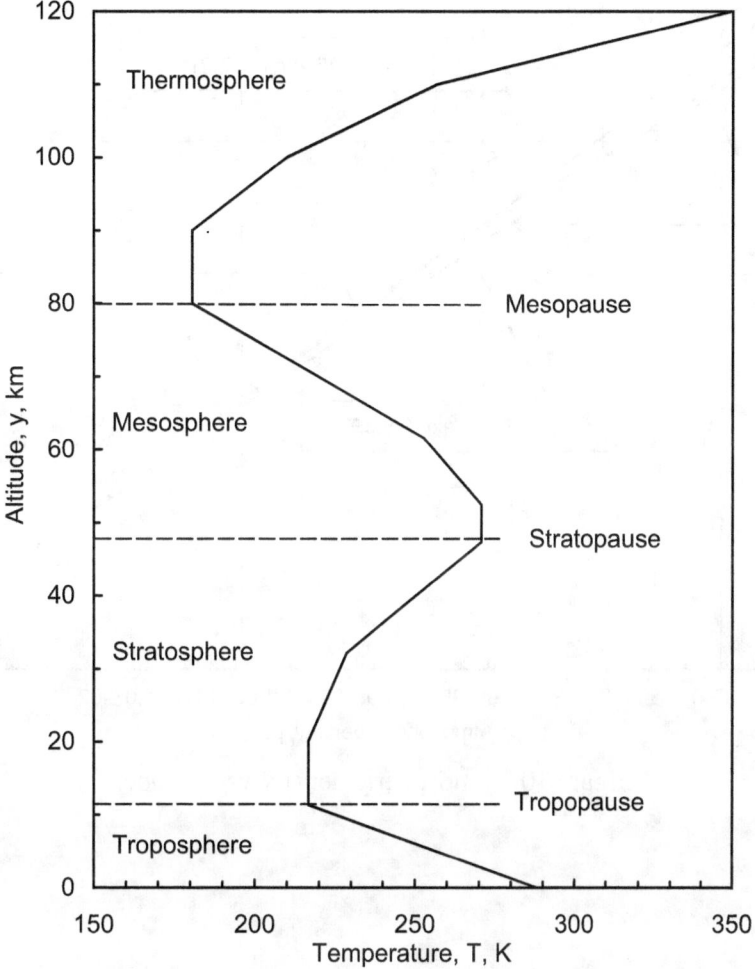

Figure 9. Temperature vs. altitude, 1976 U.S. Standard Atmosphere.

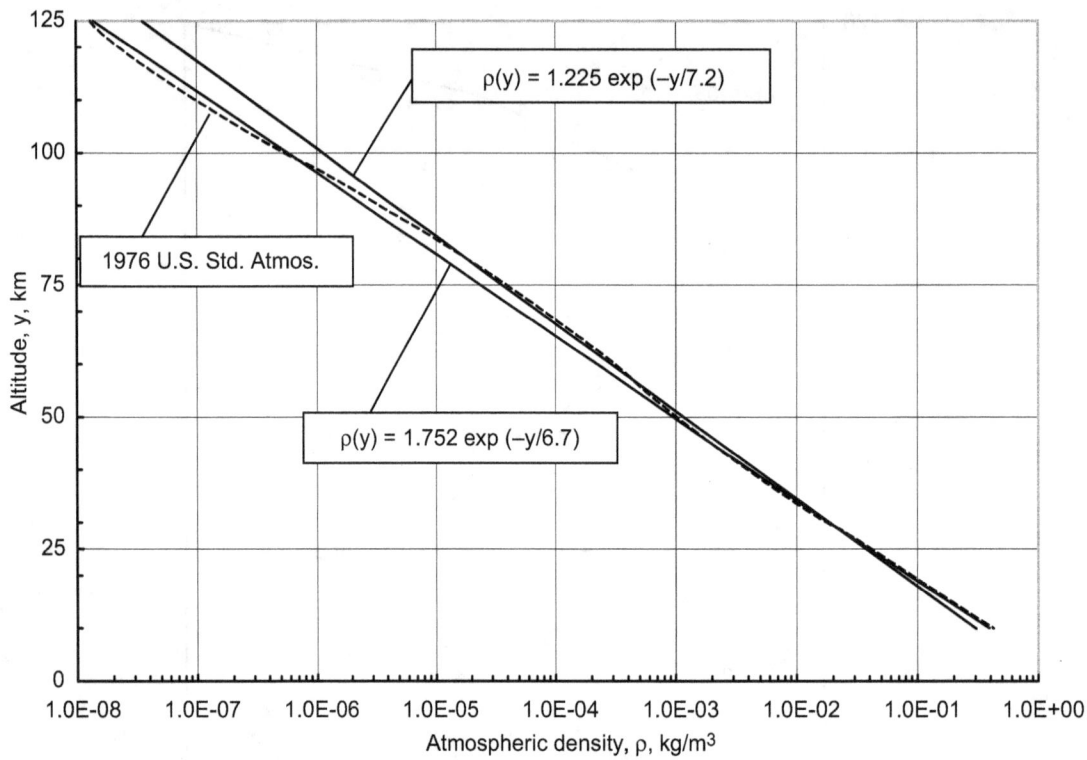

Figure 10. Atmospheric density vs. altitude.

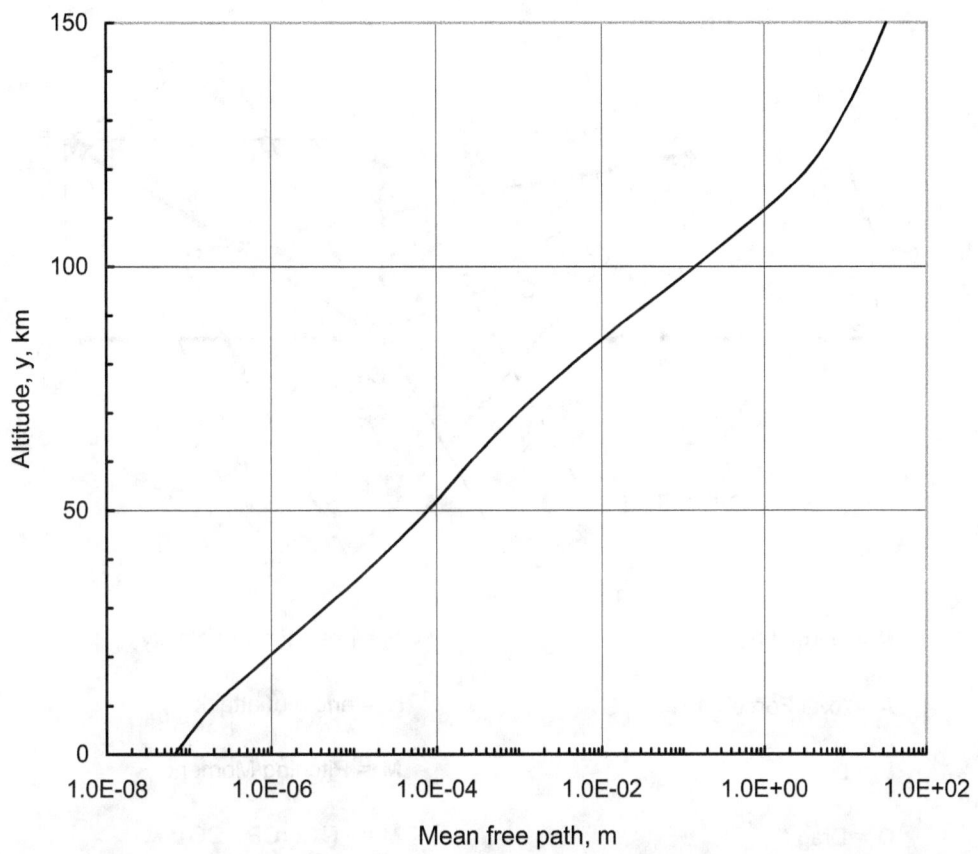

Figure 11. Mean free path vs. altitude, 1976 U.S. Standard Atmosphere.

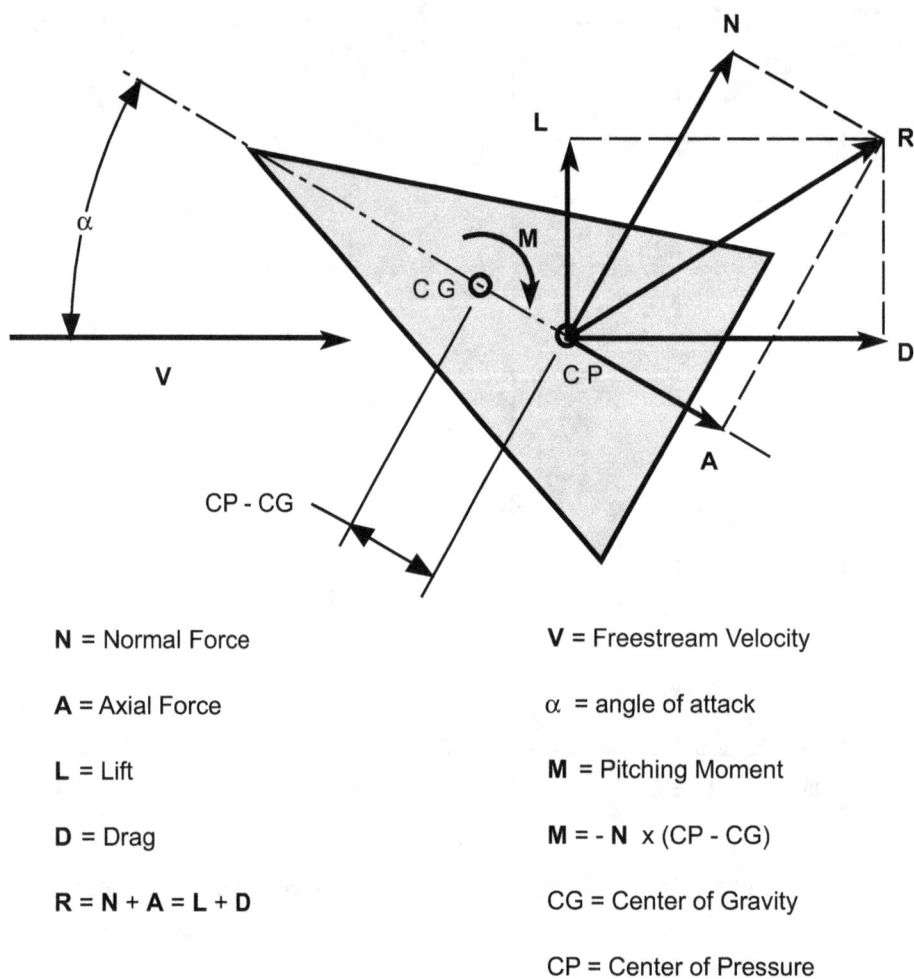

N = Normal Force   V = Freestream Velocity

A = Axial Force α = angle of attack

L = Lift   M = Pitching Moment

D = Drag   M = - N x (CP - CG)

R = N + A = L + D   CG = Center of Gravity

                              CP = Center of Pressure

Figure 12. Aerodynamic forces and pitching moment.

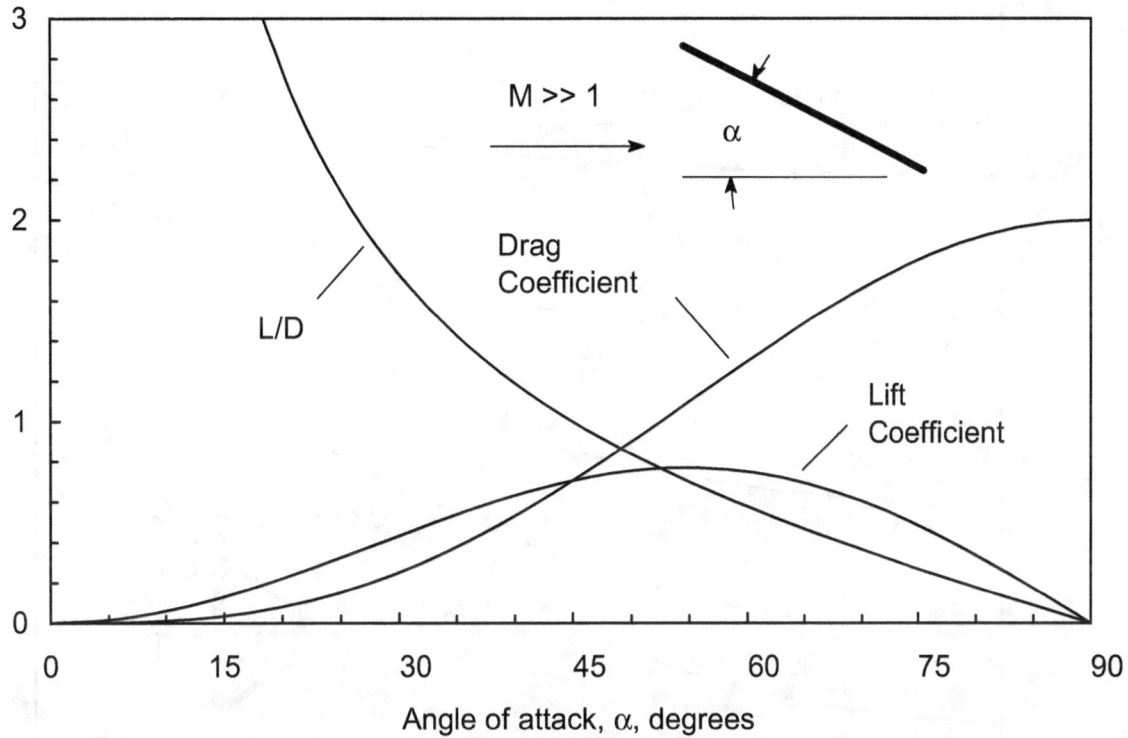

Figure 13. Hypersonic lift and drag for a flat plate at angle of attack.

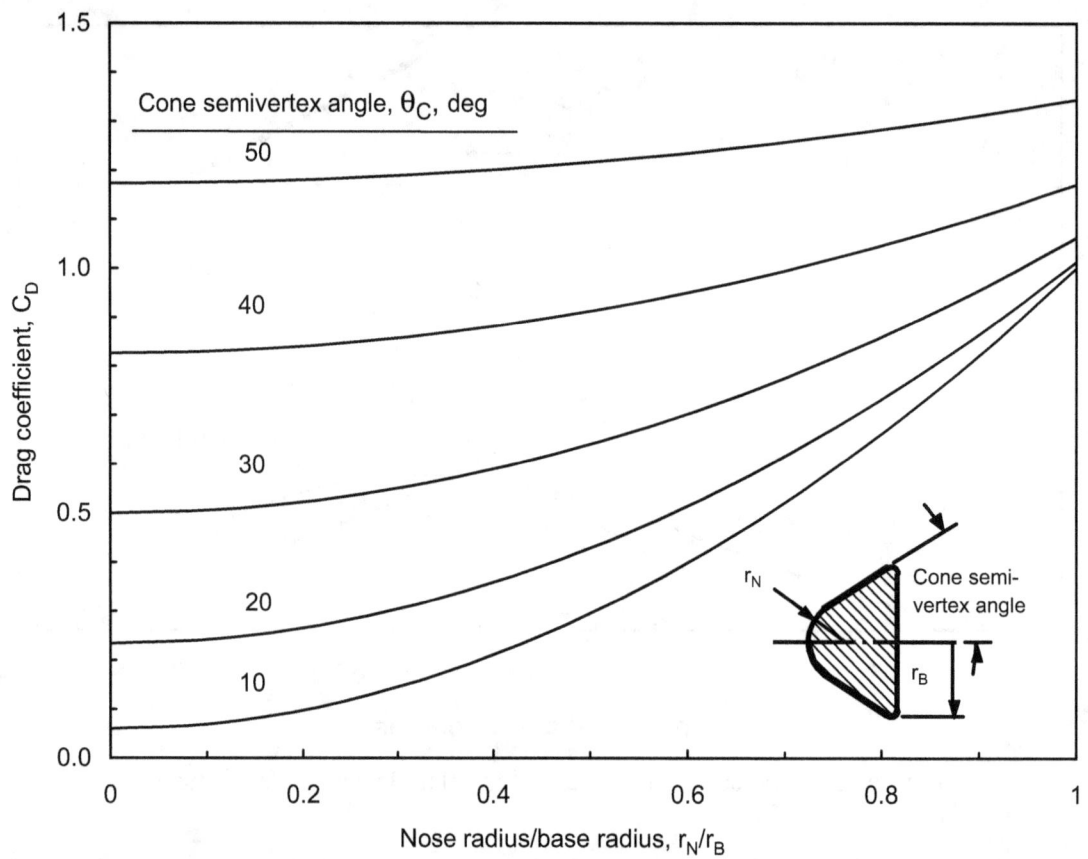

Figure 14. Hypersonic drag coefficient for sphere-cones.

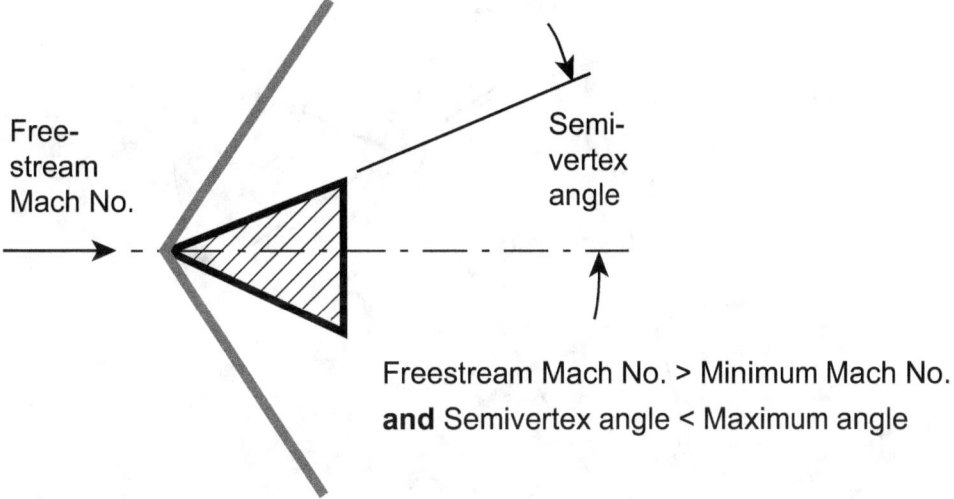

a. Sharp-nosed body with attached shock wave

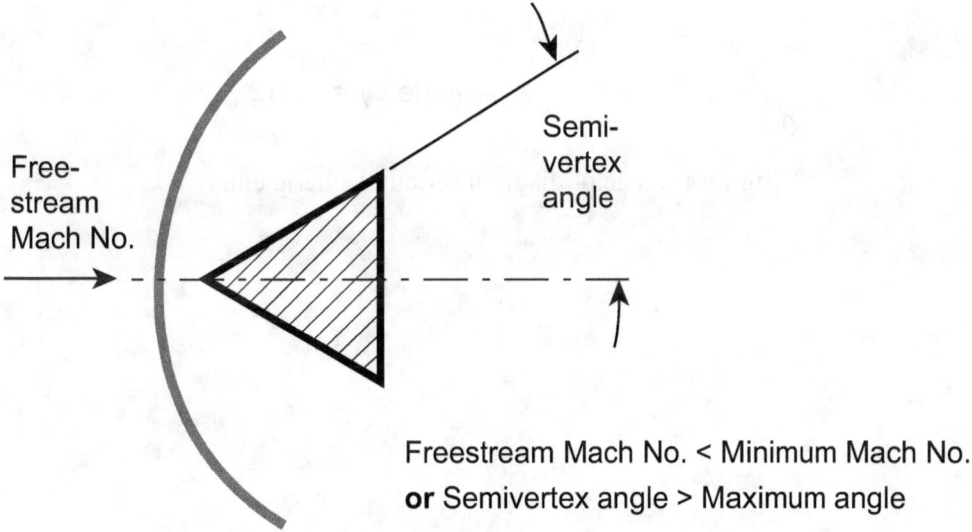

b. Sharp-nosed body with detached shock wave

Figure 15. Sharp-nosed bodies with attached and detached shock waves.

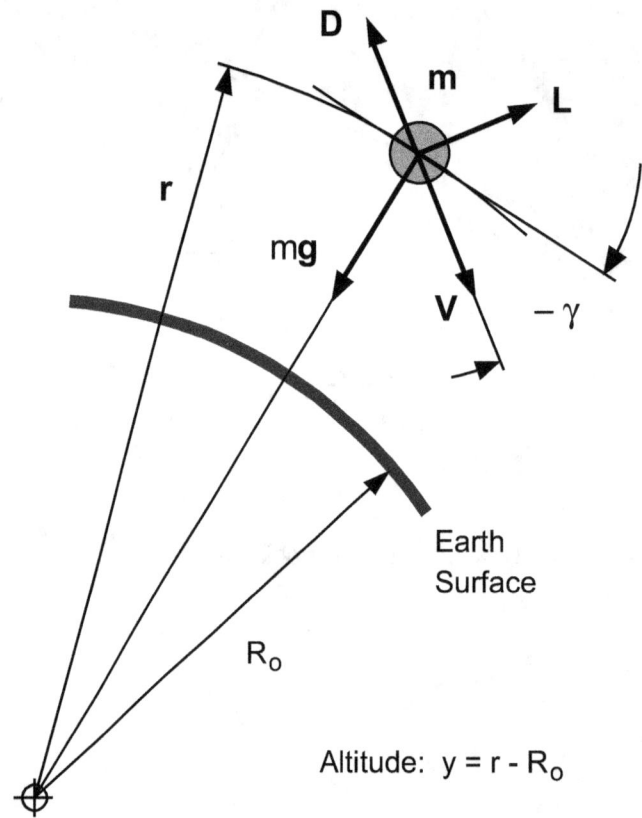

Altitude: $y = r - R_o$

Figure 16. Vector diagram for atmospheric entry.

100

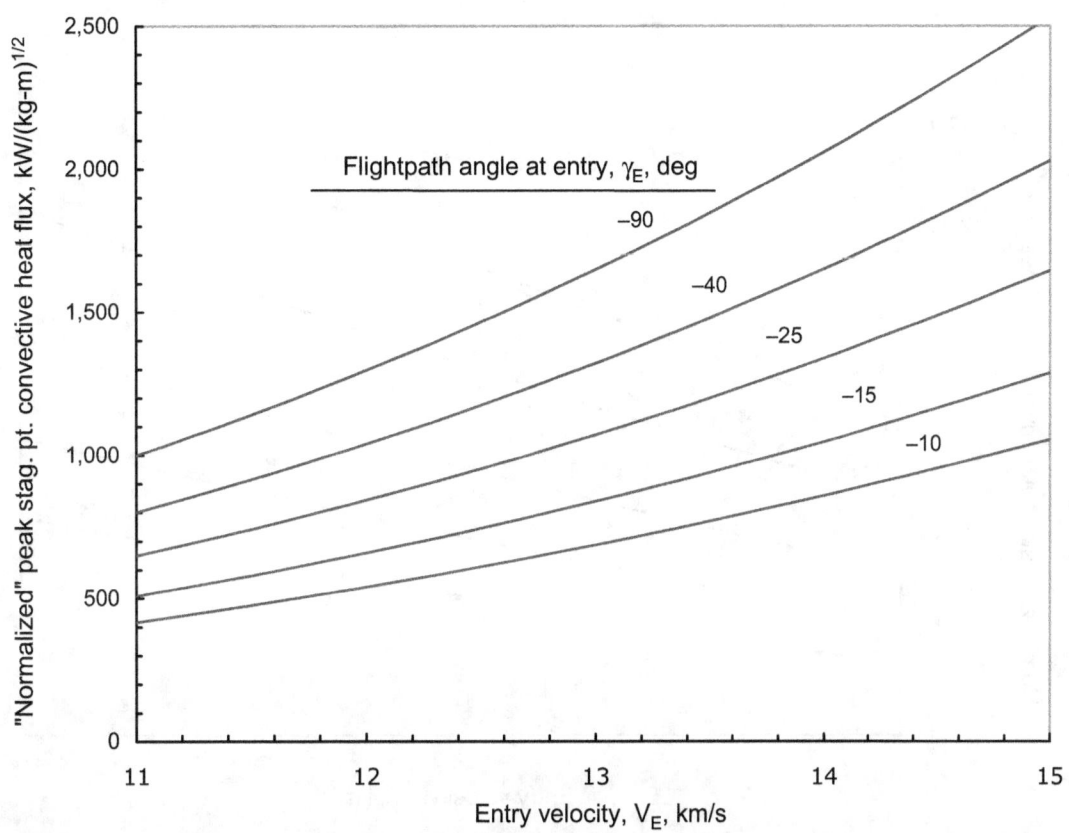

Figure 17. "Normalized" peak stagnation point convective heat flux for steep, nonlifting entry.

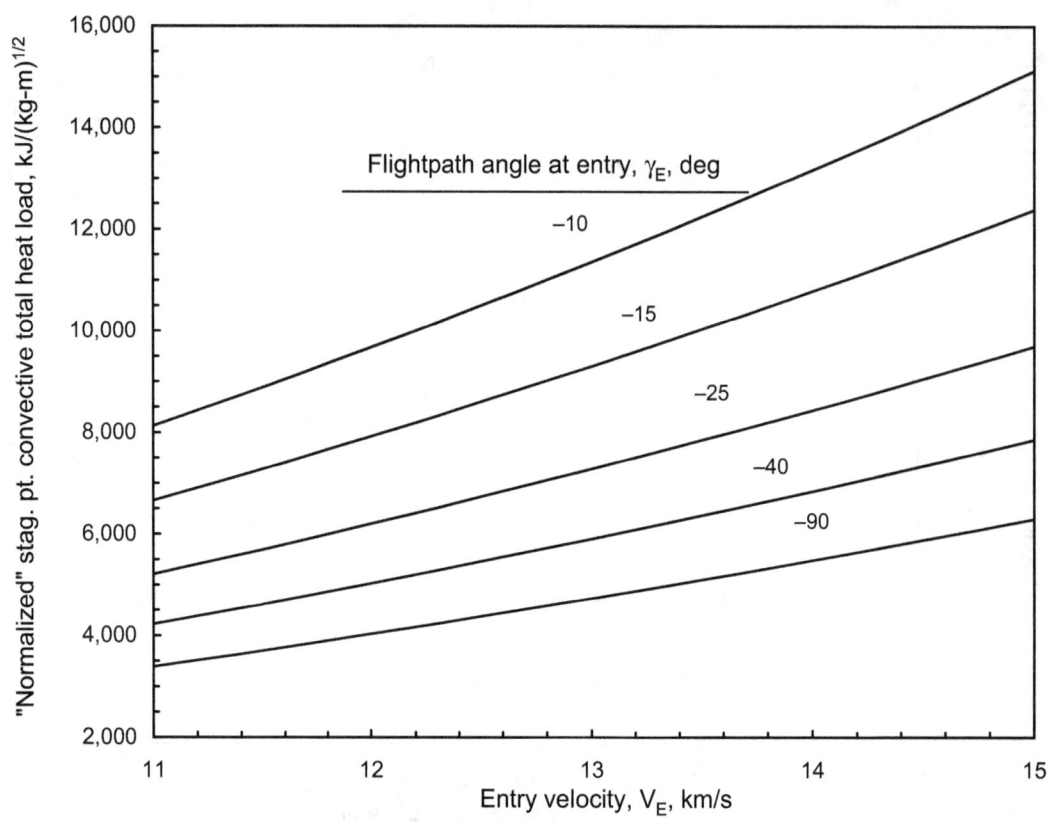

Figure 18. "Normalized" stagnation point convective total heat load for steep, nonlifting entry.

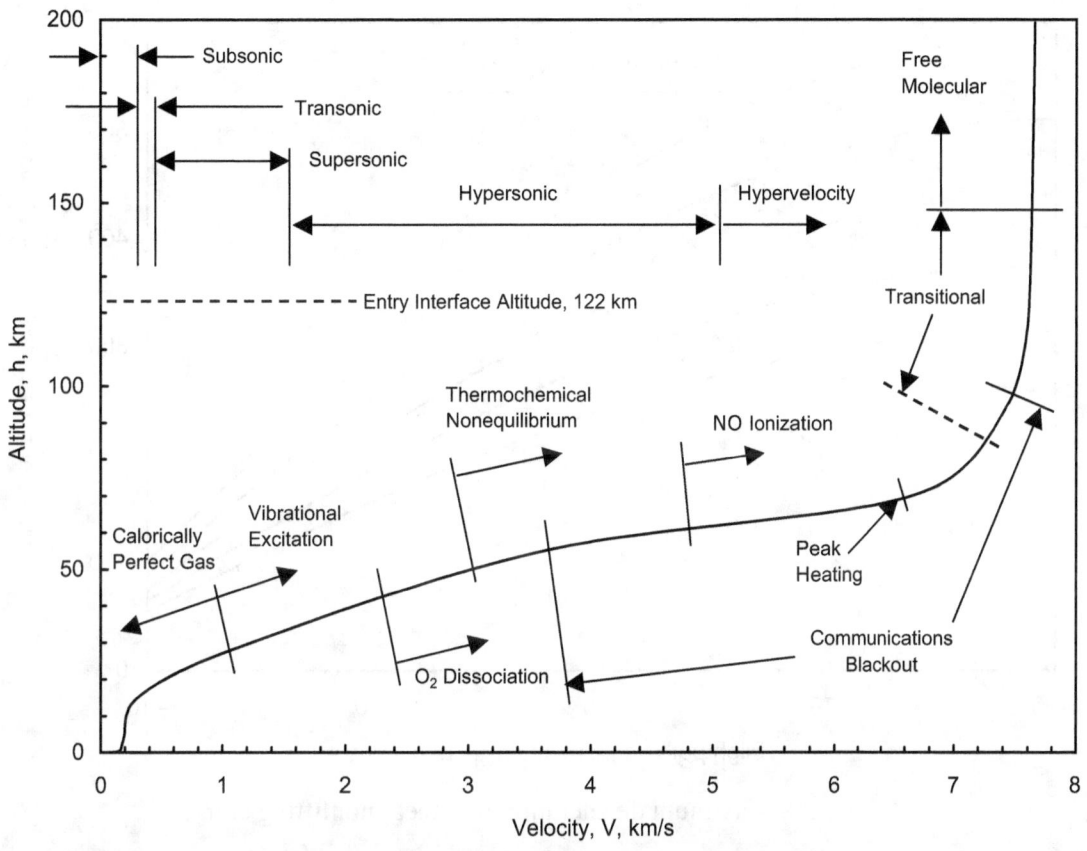

Figure 19. Space Shuttle Orbiter re-entry trajectory.

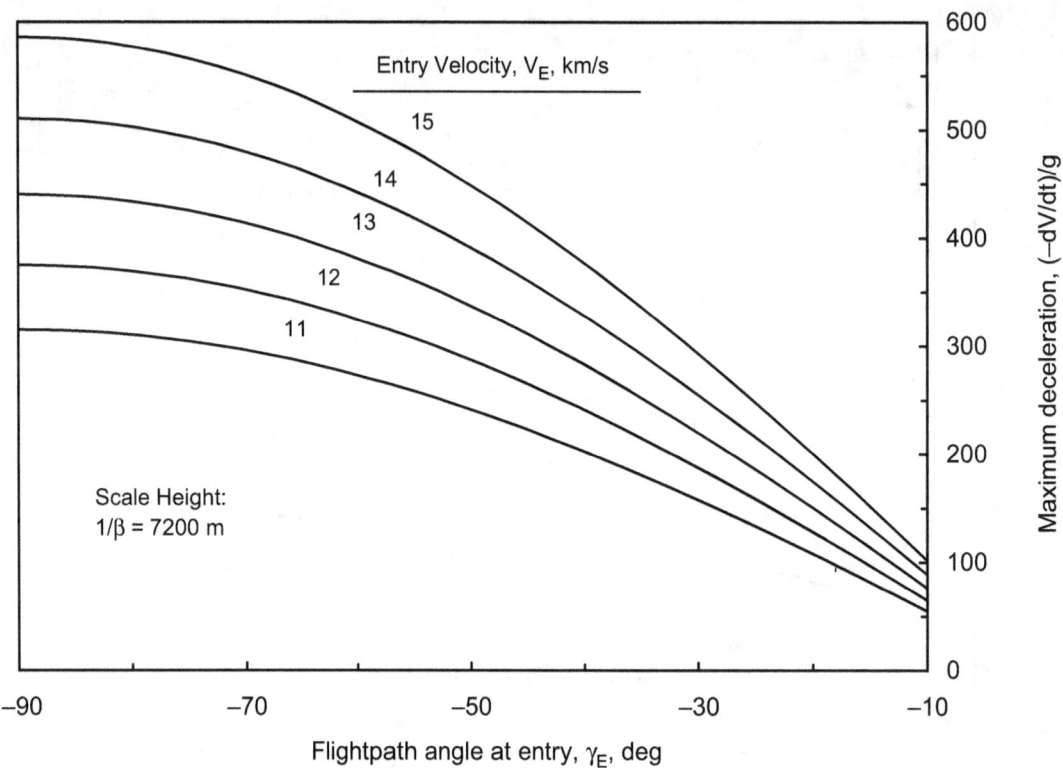

Figure 20. Maximum deceleration for steep, nonlifting entry.

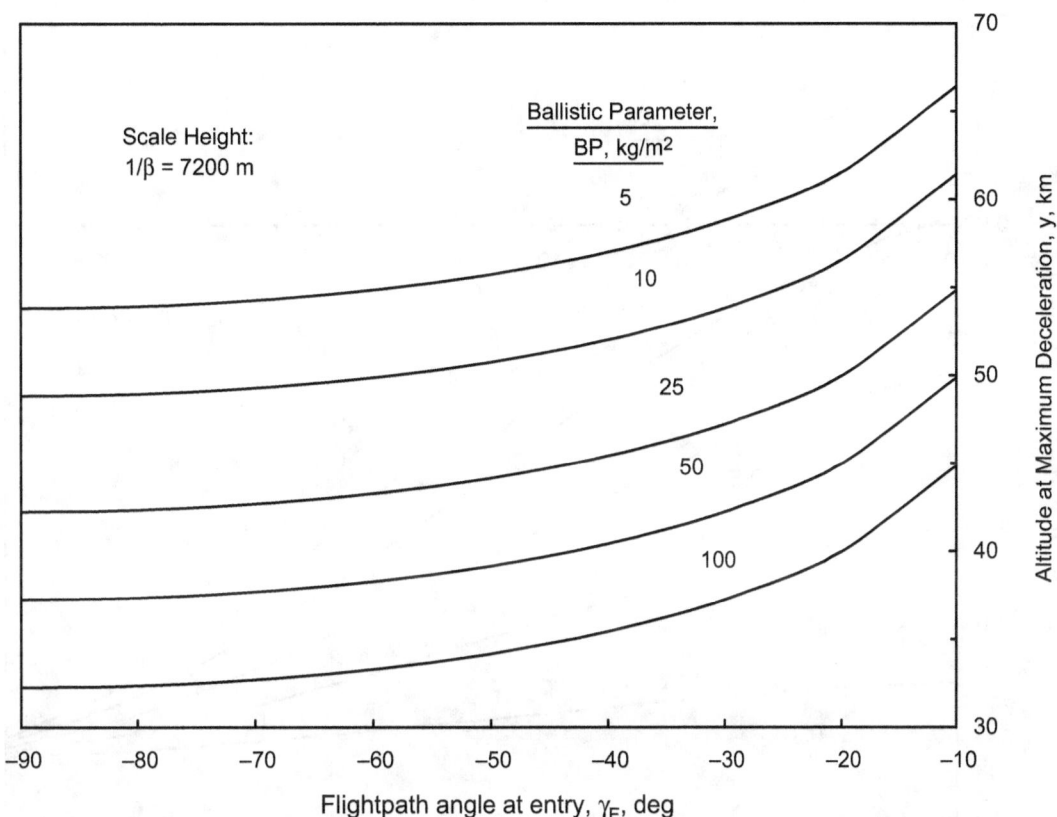

Figure 21. Altitude at maximum deceleration for steep, nonlifting entry.

105

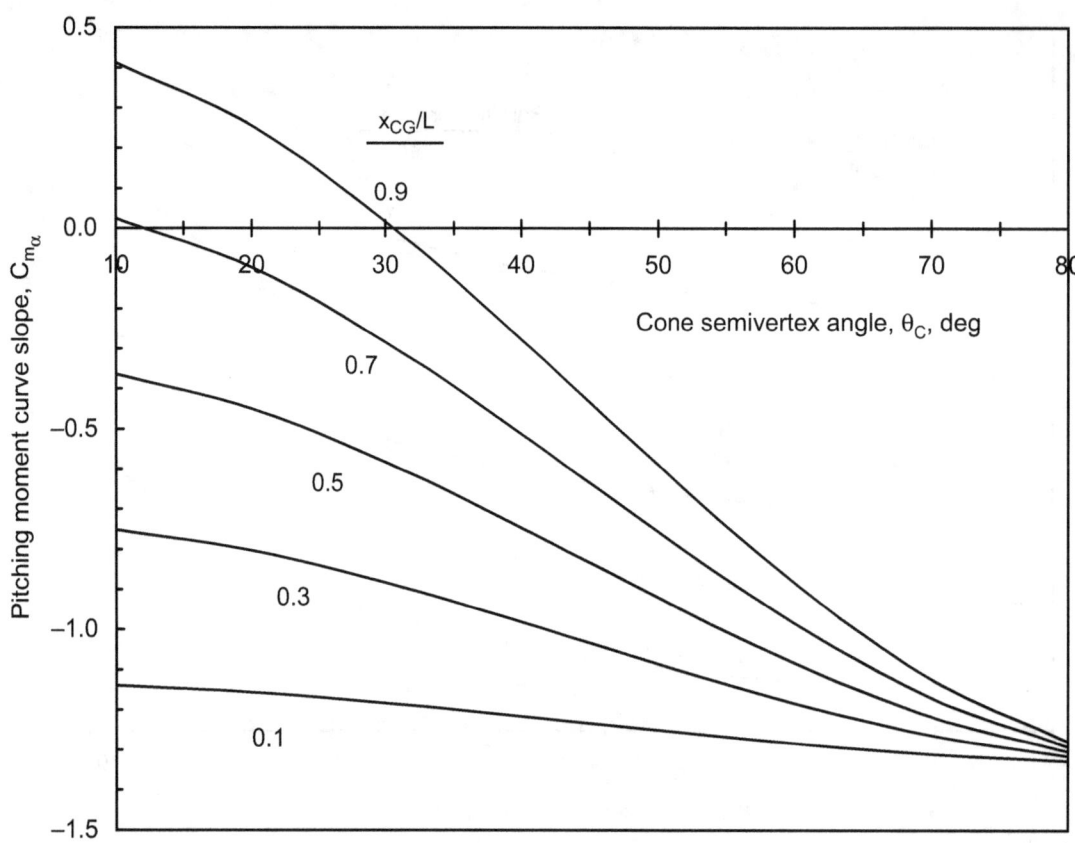

Figure 22. Hypersonic pitching moment curve slope for sharp cones.

106

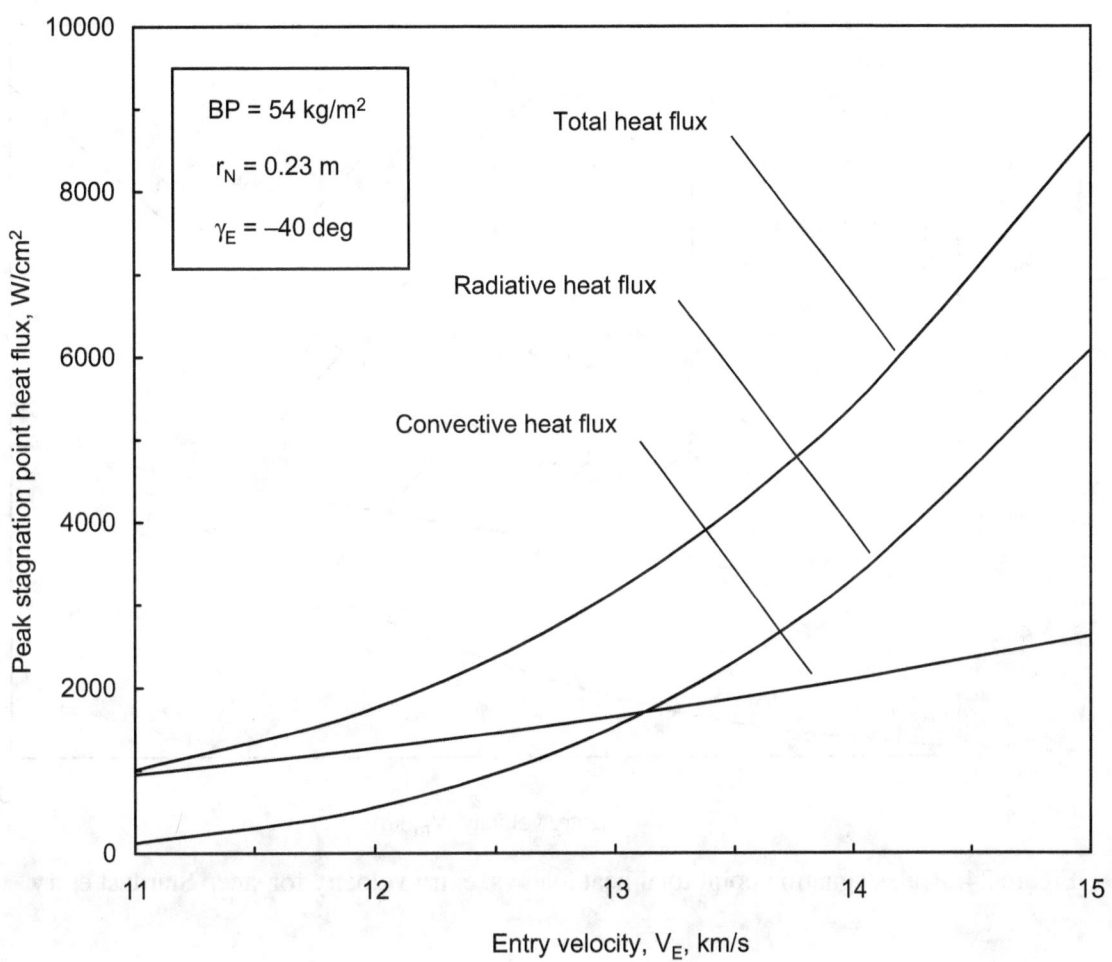

Figure 23. Peak stagnation point heat flux vs. entry velocity for steep Stardust entry.

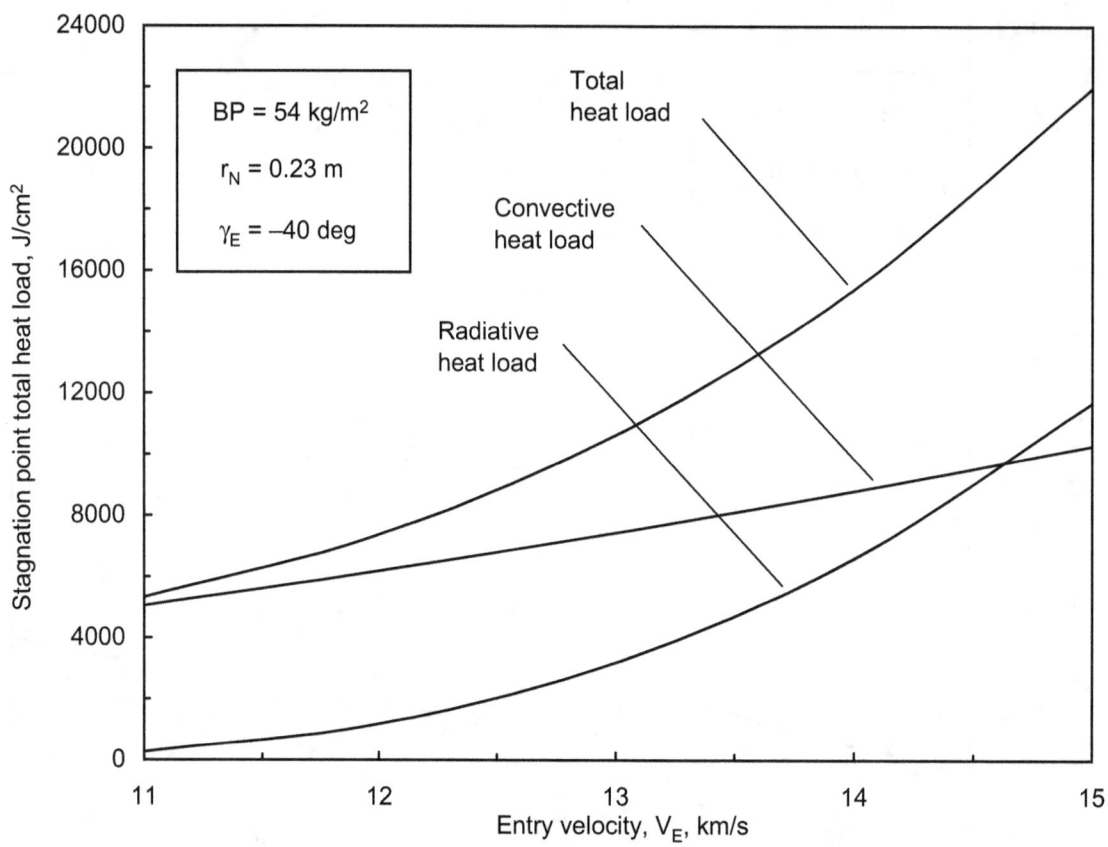

Figure 24. Peak stagnation point total heat load vs. entry velocity for steep Stardust entry.

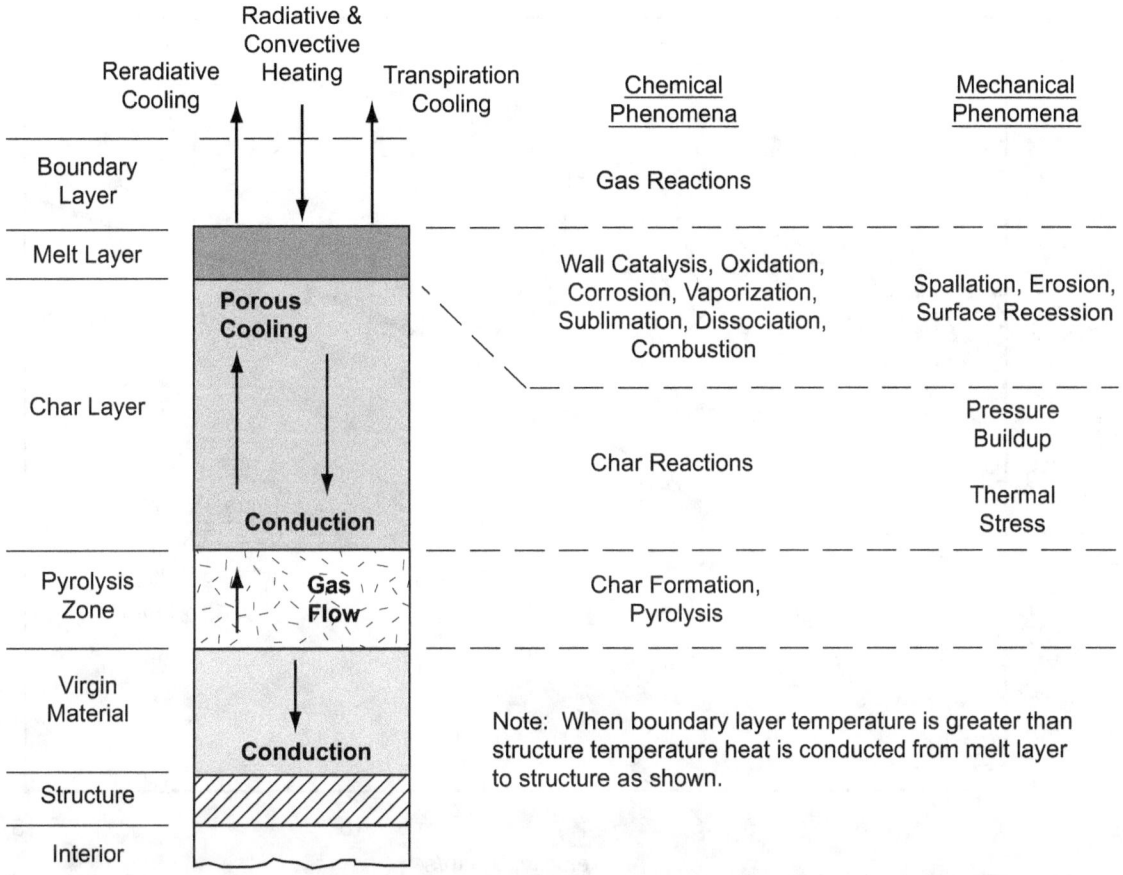

Figure 25. Schematic diagram of charring ablation process.

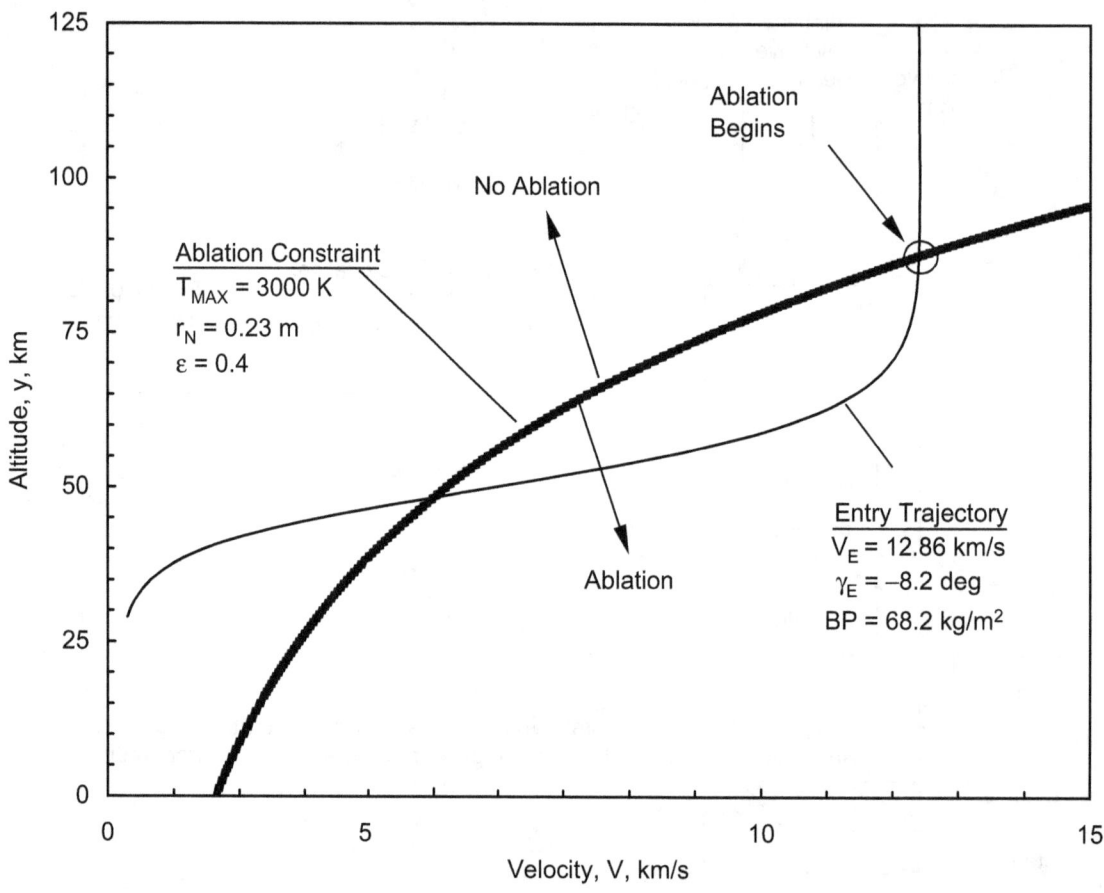

Figure 26. Atmospheric entry trajectory and ablation constraint.

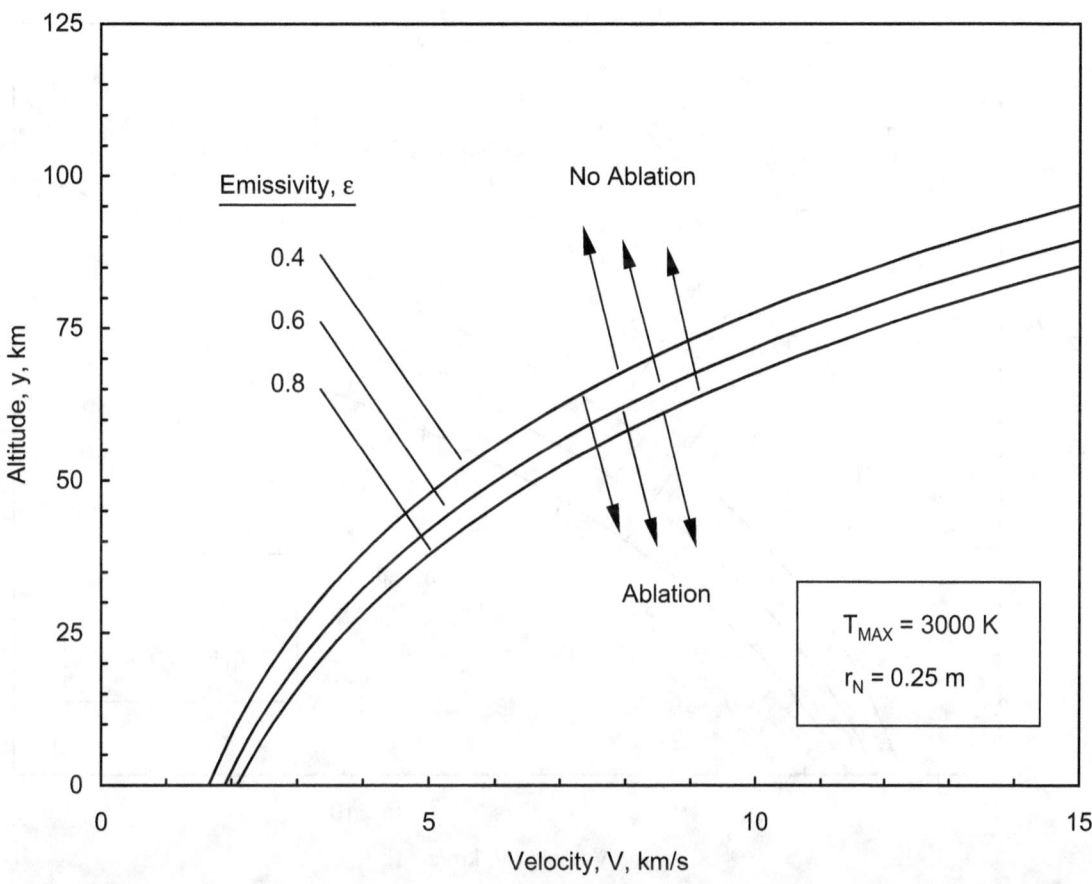

Figure 27. Effect of emissivity on ablation constraint for an axially-symmetric body.

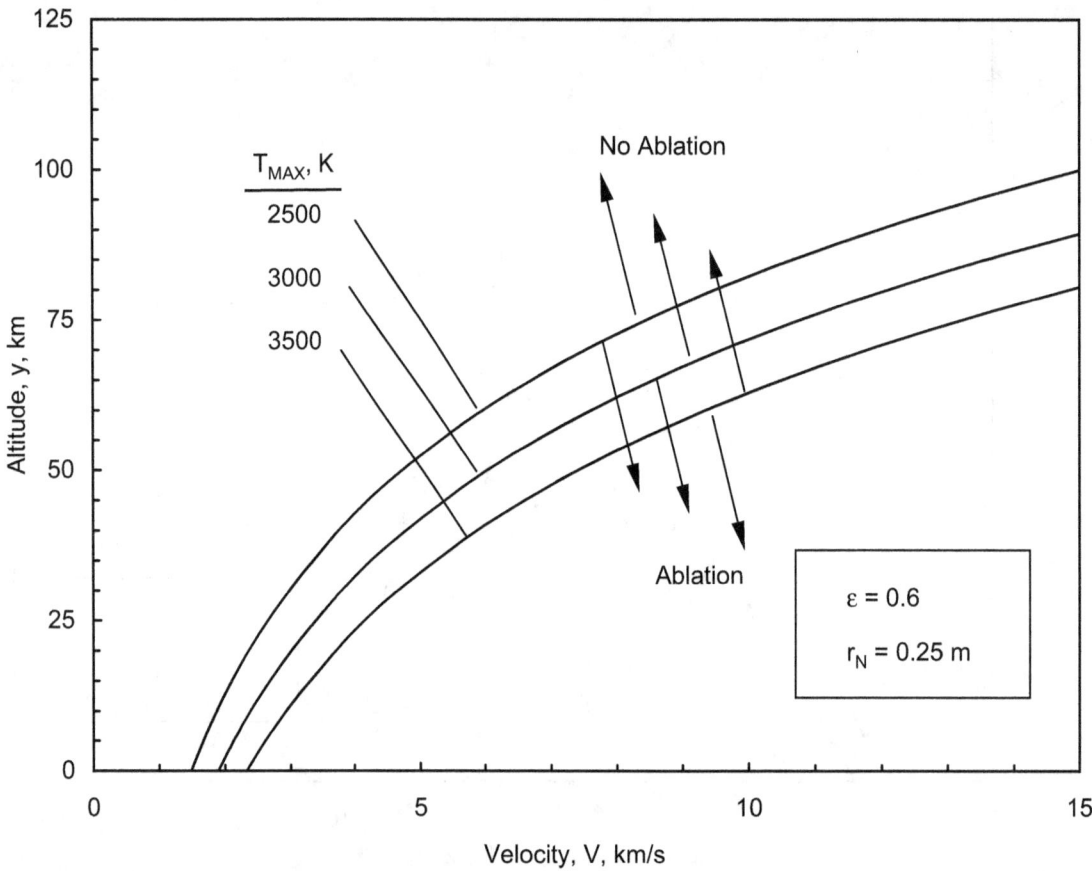

Figure 28. Effect of ablation temperature on ablation constraint for an axially-symmetric body.

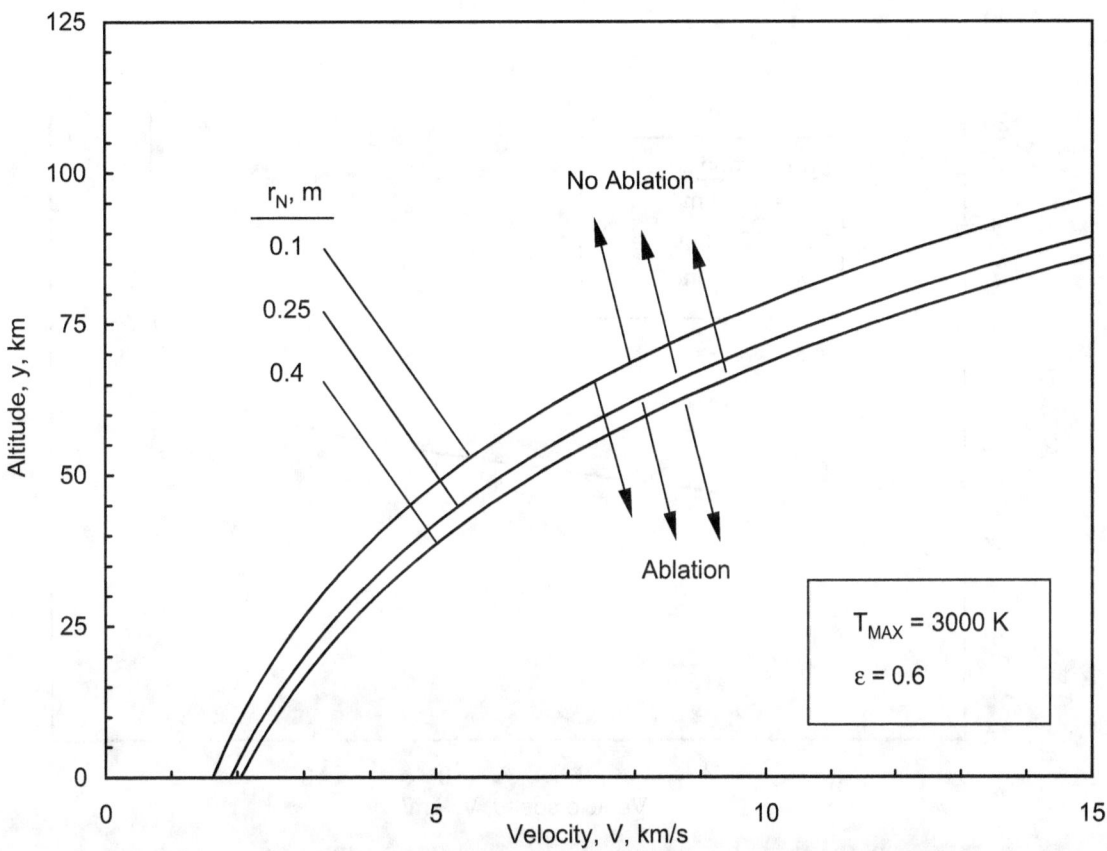

Figure 29. Effect of nose radius on ablation constraint for an axially-symmetric body.

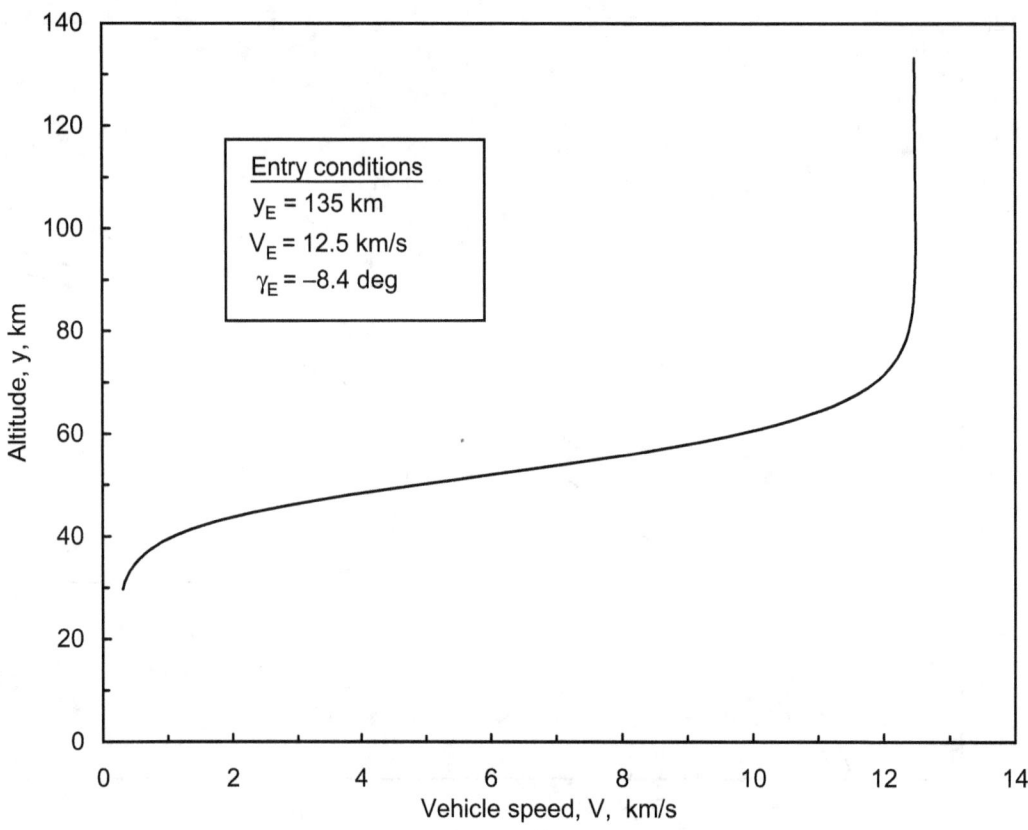

Figure 30. Stardust altitude vs. vehicle speed (NASA TRAJ Code, 1999).

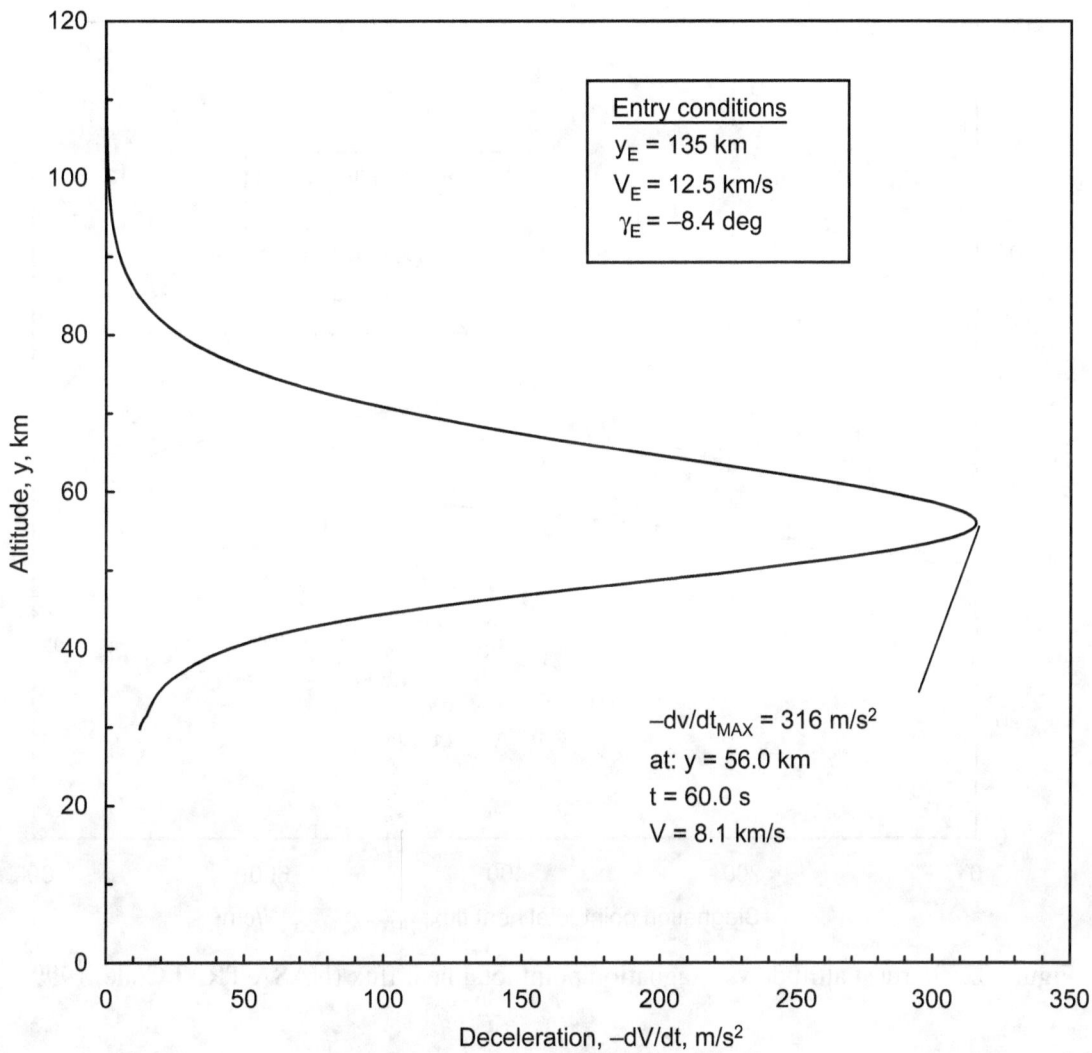

Figure 31. Stardust altitude vs. deceleration (NASA TRAJ Code, 1999).

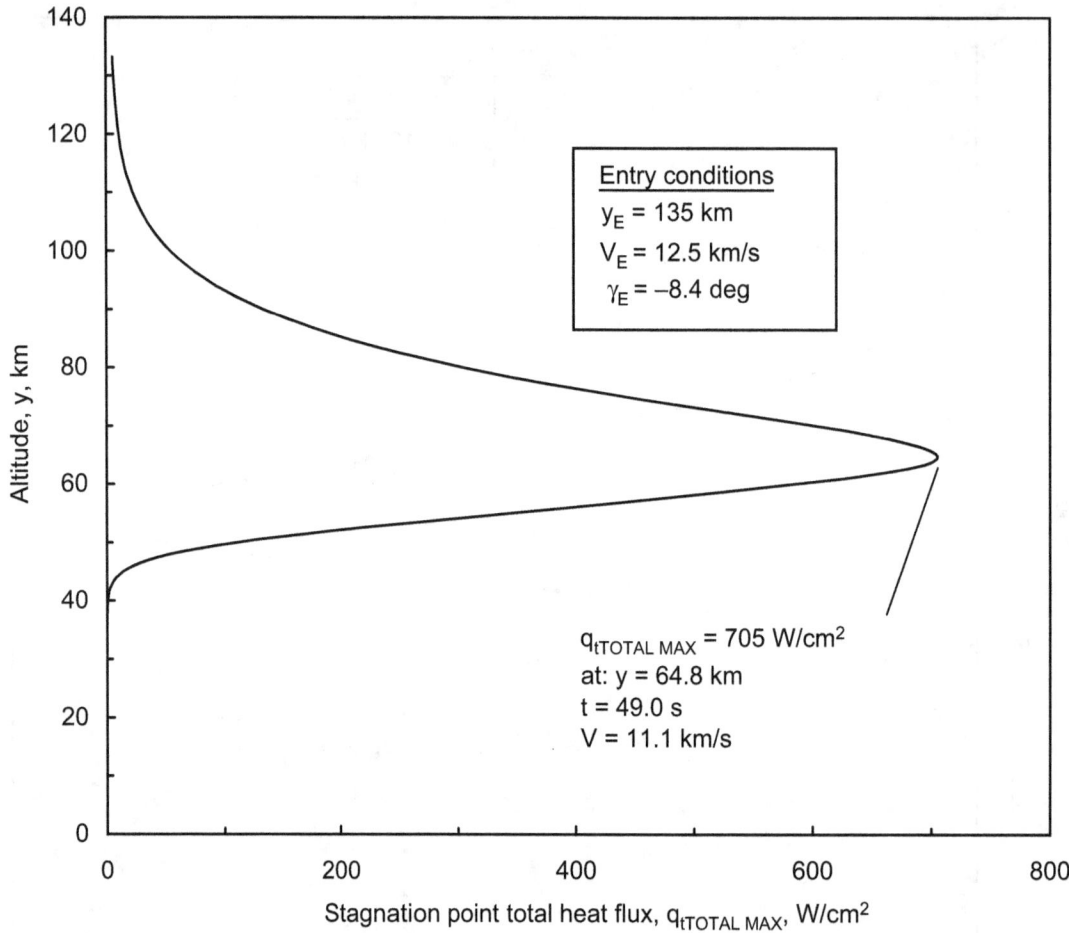

Figure 32. Stardust altitude vs. stagnation point total heat flux (NASA TRAJ Code, 1999).

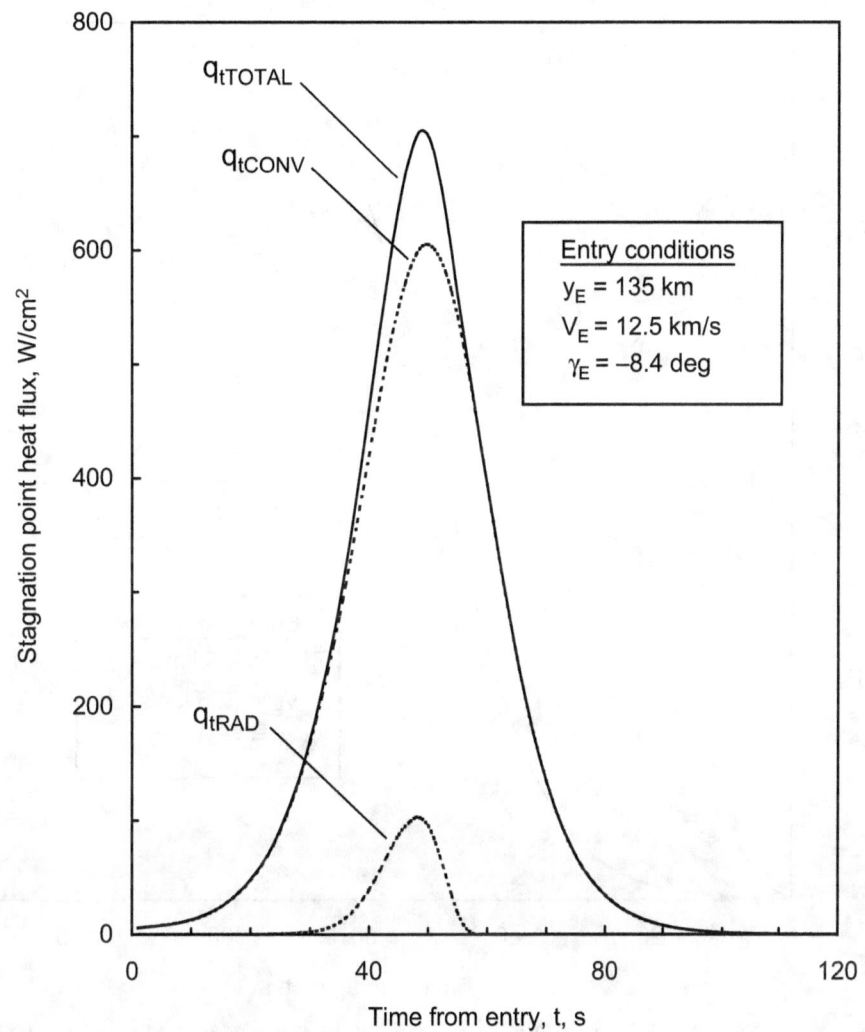

Figure 33. Stardust stagnation point heat flux vs. time from entry (NASA TRAJ Code, 1999).

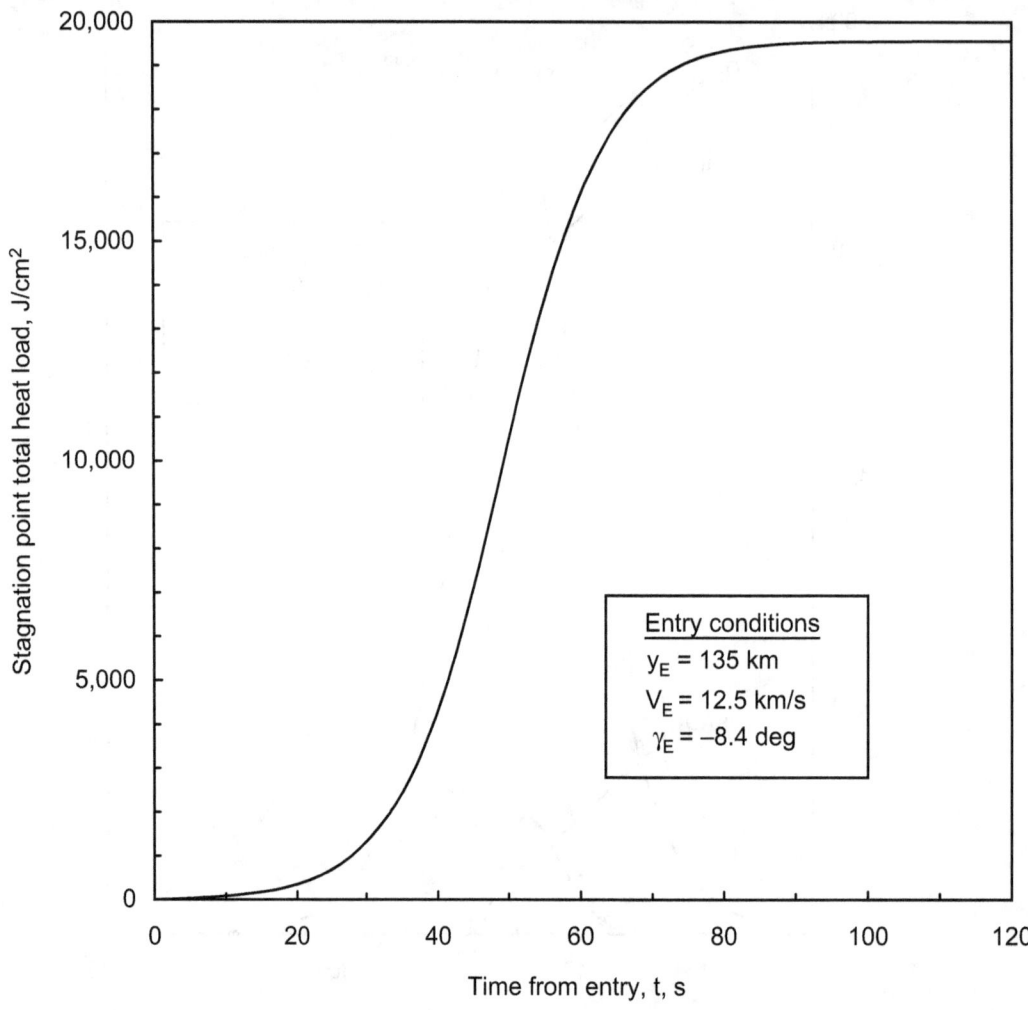

Figure 34. Stardust stagnation point total heat load vs. time from entry (NASA TRAJ Code, 1999).